BREWING IN HERTFORDSHIRE

MARTYN CORNELL

AMBERLEY

First published 2024

Amberley Publishing
The Hill, Stroud
Gloucestershire, GL5 4EP

www.amberley-books.com

All images are from the author's collection, unless otherwise stated.

ISBN 978 1 3981 0844 8 (print)
ISBN 978 1 3981 0845 5 (ebook)

British Library Cataloguing in Publication Data.
A catalogue record for this book is available from the British Library.

Typesetting by SJmagic DESIGN SERVICES, India.
Printed in the UK.

Contents

Introduction 4
Ashwell 6
Baldock 10
Barkway 14
Barnet 14
Berkhamsted 16
Bishop's Stortford 17
Borehamwood 20
Cheshunt 21
Furneux Pelham 22
Hadley 24
Harpenden 27
Hatfield 29
Hemel Hempstead 31
Hertford 33
Hitchin 43
Hoddesdon 46
Ickleford 49
Kimpton 49
Kings Langley 50
Kings Walden 51
Little Hadham 53
London Colney 54
Markyate 54
Puckeridge 55
Redbourn 56
Rickmansworth 57
Royston 60
St Albans 62
Sandon 66
Sawbridgeworth 67
Shenley 68
Standon 68
Stanstead Abbotts 68
Stapleford 69
Stevenage 69
Tring 71
Walkern 74
Waltham Cross 76
Ware 76
Watford 79
Welwyn 86
Wheathampstead 87
Whitwell 88
Appendix: Hertfordshire Breweries Founded Post-1980 89

Introduction

For centuries Hertfordshire was known for malt, and malting was the county's first industry. John Taylor wrote in 1636: 'Hertfordshire is a County that surpasseth all other Countries and Counties for making the best malt.' Almost a century later, in 1724, Daniel Defoe wrote that Hertfordshire malt was 'esteemed the best in England'. The northern and eastern parts of the county, with their light, well-drained, chalky souls, grew barley that was well-suited to turning into malt. Hertfordshire was also on the route from other rich barley-growing areas – Bedfordshire, Cambridgeshire, Suffolk and Norfolk – down Ermine Street to London.

By 1339 there were complaints of severe damage to the roads leading into and out of Ware because of the 'unreasonable' loads of malt travelling through the town. The first turnpike road in England was set up at Wadesmill on Ermine Street, leading into Ware, in 1663 specifically to pay for the damage the grain wagons were causing to the road. Eventually, travellers wanting to avoid a route through Royston and Ware rutted up by grain wagons turned off the old North Road at Huntingdon and came south via Biggleswade, Baldock, Stevenage and Welwyn, creating the Great North Road.

Ware was particularly placed to become a centre of the malting trade, being where Ermine Street meets the navigable River Lea. Barley could be brought down to the town, malted and then loaded onto barges for the journey into London. In 1788 Ware had thirty-three maltings. By the end of the eighteenth century 5,000 quarters of malt a week, around 700 tons, were travelling from Ware to London, enough to produce more than 60 per cent of the capital's beer.

Almost all Hertfordshire brewers had their own maltings attached to their breweries, and a directory of 1832 lists eighty-two maltsters in Hertfordshire, right across the county. By now Ware had eighty maltings in use. A maltings is a very distinctive building: long and narrow, with two or more storeys containing the floors where the soaked barley is spread to start sprouting, a process that begins the conversion of the starch in the barley seed into the sugar that the brewer needs to make beer from, and at one end the tall kiln where the green malt is dried.

The brewing historian Peter Mathias wrote in 1959: 'The tapering flues of the malt kilns gave as characteristic an appearance to the little Hertfordshire malting towns as did the oasts to a Kentish hop village.'

However, by the 1980s there were only four malting firms left in Hertfordshire, and today there is just one, French & Jupps of Stanstead Abbotts, which specialises in the brown and coloured malts that Hertfordshire became famous for in the heyday of porter brewing. There are plenty of reminders of the county's malting history surviving, however, with some eighty maltings still standing and being used for a wide variety of different activities, including schoolrooms in Stevenage, an arts centre in Ware, a library in Baldock, flats in Hertford and a church hall in Kings Langley.

Hertfordshire also only has one survivor from its 140 or so pre-twentieth-century brewers: McMullen's, in the county town. But a revival of craft brewing since 1981 has seen dozens of start-ups add their stories to Hertfordshire's brewing history.

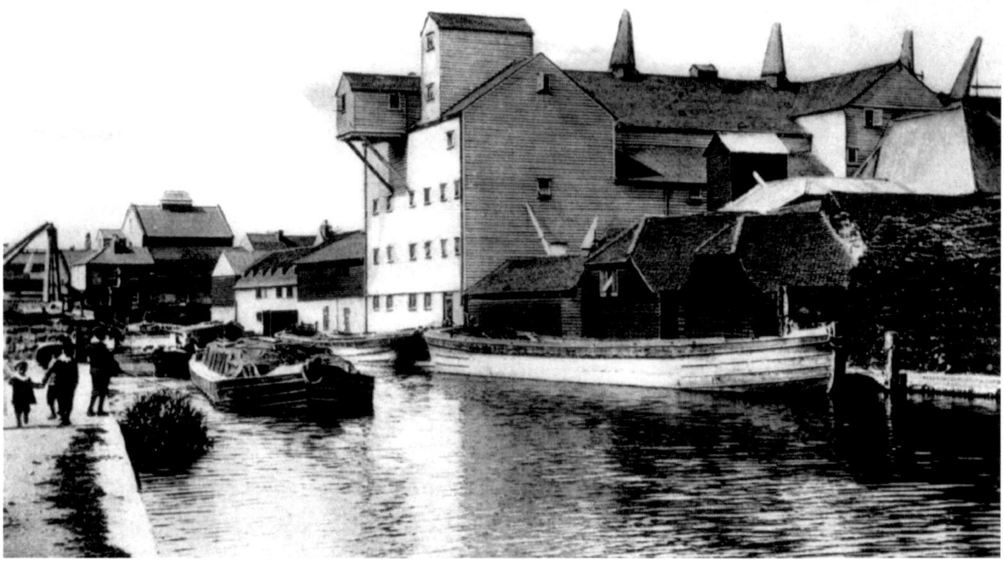

Maltings on the River Lea in Ware, with malt barges ready for the journey to London, *c.* 1905.

ASHWELL

Fordham's

The Fordhams were an old-established Hertfordshire family, living in the county since the early sixteenth century. One, John Fordham, ran a brewery in Hertford with his brother-in-law, Richard Flower, in the 1780s. His nephew, Edward George Fordham, was living at Ashwell Bury with his wife by 1810.

Soon after the 1830 Beer House Act a Mr Porter had commenced brewing in premises at the corner of Mill Street, Ashwell, 'by Carter's pond'. In around 1832, Porter's plant was purchased by Edward George Fordham's son, Edward King Fordham, then twenty-two, and his younger brother, Oswald, and moved to Bury Farm.

The railway came to Ashwell in 1851, and Fordham's took advantage of the opportunities the railway offered, building up a chain of ten 'Railway Inns' and 'Station Hotels' from Ashwell to Hornsey in North London.

A new brewery 'on a larger scale' was opened in 1860. Oswald Fordham died in 1862, and was replaced in the partnership by a cousin, Herbert Fordham. Edward King Fordham died in March 1889, aged seventy-nine. His share in the brewery went to his second son, Wolverley Attwood Fordham, born in 1859, a polo player and big-game hunter in Africa and America.

In 1893, the brewery's annual outing was to Hastings. Among the guests on the trip was George Ruddle, who became the manager of Henry Parry's brewery in Langham, Rutland, in 1896. Ruddle married Herbert Fordham's youngest daughter, Nora, and in 1911, after Parry died, he bought the Langham Brewery, which continued under the Ruddle name.

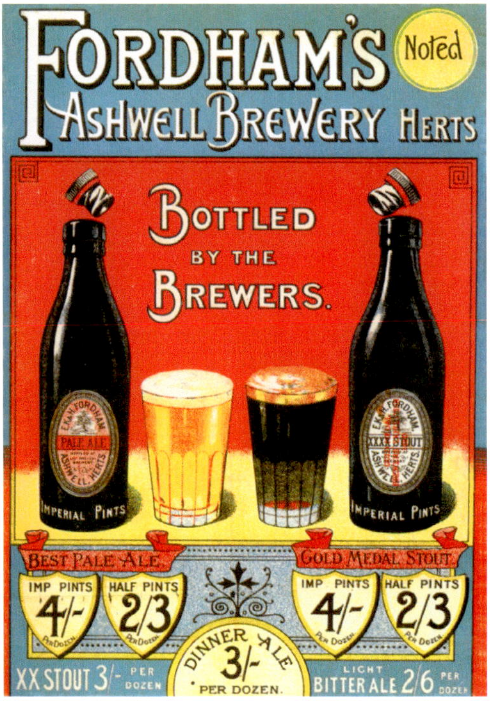

A showcard from before 1897, when Fordham's became a limited company, promoting the brewery's screw-top bottles.

In the 1890s the Ashwell Brewery was producing a typical range of beers, eight in all, including XXX mild and AK light bitter ale at 1*s* a gallon, XXV old ale and OO Imperial Ale at 1*s* 8*d* a gallon. Farmers could buy 'brewers' grains' (the left-over malt after mashing) from the brewery for 4*d* a bushel for animal feed. Anyone who turned up at the brewery with a bowl and a halfpenny could buy the fresh, foaming 'barm' or yeast for breadmaking.

In April 1897 the partnership at the brewery was turned into a limited company, E. K. and H. Fordham Ltd. There were 103 freehold tied houses, fifteen copyhold, six on long leases and seven on short leases, making 131 in all.

Just over twelve months later, in April 1898, Fordham's took over the Hyde Hall Brewery near Sandon, the last commercial farm brewery in Hertfordshire. Fordham's leased it for the extra brewing capacity it provided. But in 1899 the Hyde Hall Brewery was closed, and the fermenting plant at Ashwell increased by 400 barrels to compensate.

The Ashwell Brewery itself still had an operating farm alongside. The farm received much of the blame when the London brewing chemist Lawrence Briant was called in in 1900 to try to cure an outbreak of ropeyness (a bacterial infection) and fretfulness in Fordham's beers. Briant's recommendation was to completely clear away the farm buildings. If that was not possible, he said, Fordham's must throw away the old open cooler and install a closed one, throw away the old mash tun, keep up a high hopping rate and check their yeasts regularly.

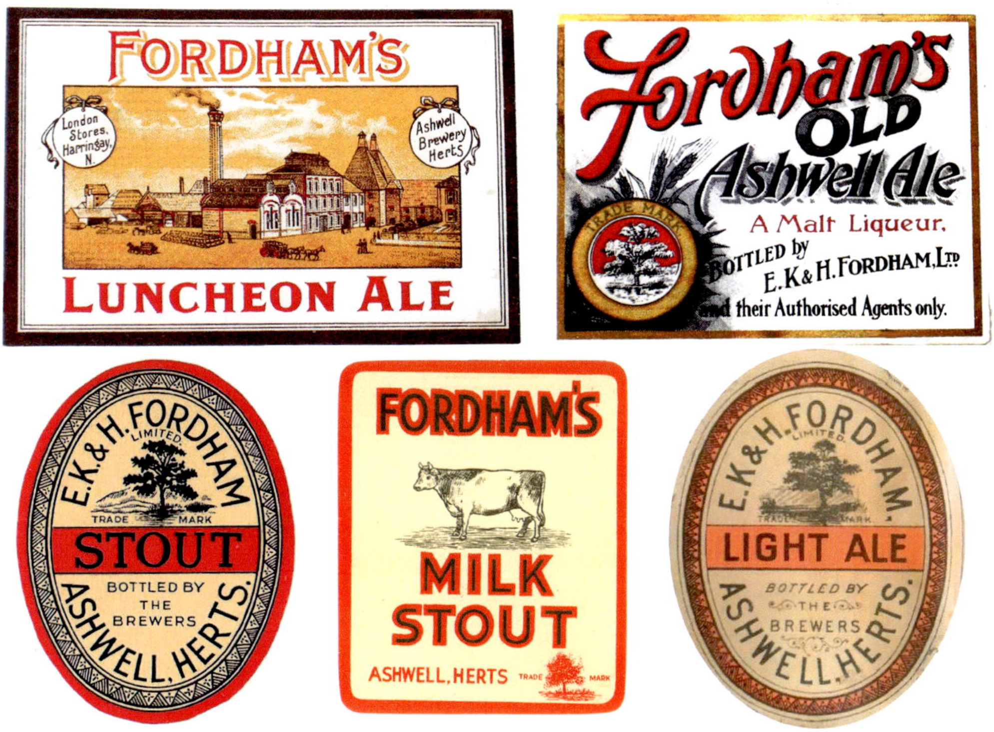

Five Fordham's bottle labels, showing the company's 'ash and well' trademark, while the Luncheon Ale label depicts the brewery and the maltings alongside.

In the 1930s the average age of the Fordham's board of directors was sixty-seven. However, there was young blood about: Anthony Eustace Hill, aged twenty-five, a grandson of Edward King Fordham, who joined the board just before the start of the Second World War. With the outbreak of war, Anthony became one of the RAF's leading photographic reconnaissance experts, which involved flying unarmed Spitfire aircraft fitted with cameras over top-secret German installations in occupied Europe. On 21 October 1942 he was shot down while on a mission, and died a short time later.

After the war, most family shareholders wanted the sale of Fordham's to another brewer. In September 1952 the expansionist firm of J. W. Green in Luton snapped up Fordham's and its ninety-one surviving pubs. The site continued as a bottling plant until 1965, with the buildings demolished in 1973.

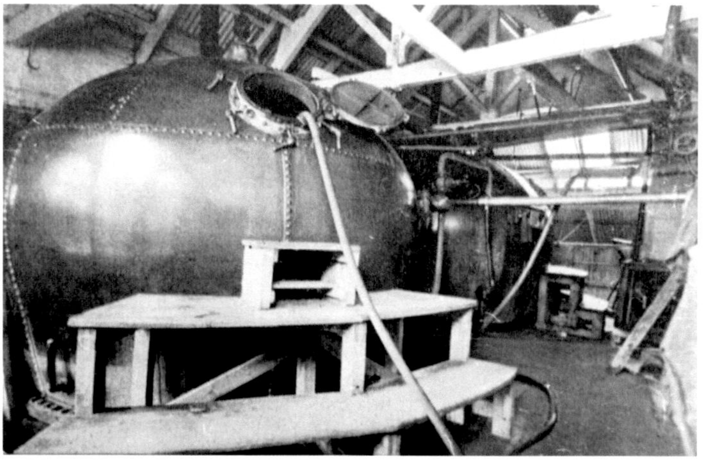

Left: Brewing coppers at the Ashwell Brewery, 1932. (Courtesy of Ashwell Museum)

Below: Fordham's directors in the boardroom in 1932. Wolverley Fordham is on the far left. (Courtesy of Ashwell Museum)

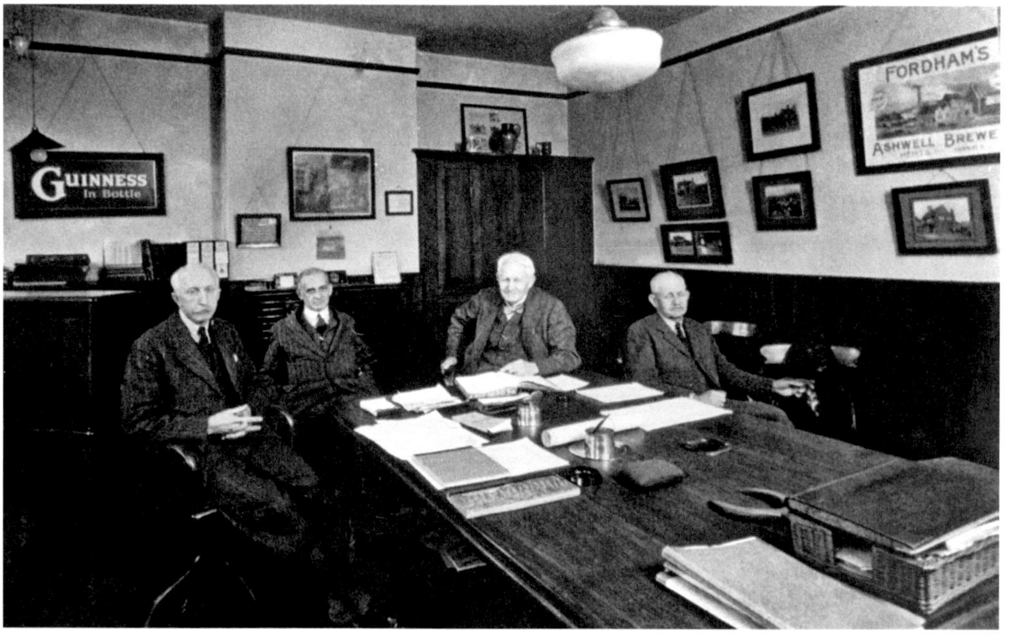

Westbury Brewery

Benjamin Christy and John Sale were neighbouring farmers and maltsters: Christy at Westbury Farm, Sale at Farrow's Farm. Their joint brewery, doubtless springing out of their farm brewing operations, was in operation by 1843, when Isaac Picking, carpenter, and Peter Bailey, bricklayer, were paid for work at 'Christy and Sale's new brewery'.

John Sale died in January 1862, aged sixty-four, at Farrow's Farm, and Christy carried on the business on his own. By the time of the 1866 *Kelly's Directory*, his son Benjamin Christy junior was in charge.

Benjamin Christy junior carried on brewing until at least 1878. In around April 1879, however, the brewery changed hands. The new owner was Joshua Richmond Page, a Baldock maltster. He had probably retired from running the business by 1898, when the concern was described by *Kelly's* as Page and Co. and was being run by Page's son, Joshua junior.

In spring 1921 the brewery's twenty-five pubs, plus an off-licence at Mangrove Green, near Luton, were sold to Wells and Winch of Biggleswade for £9,730.

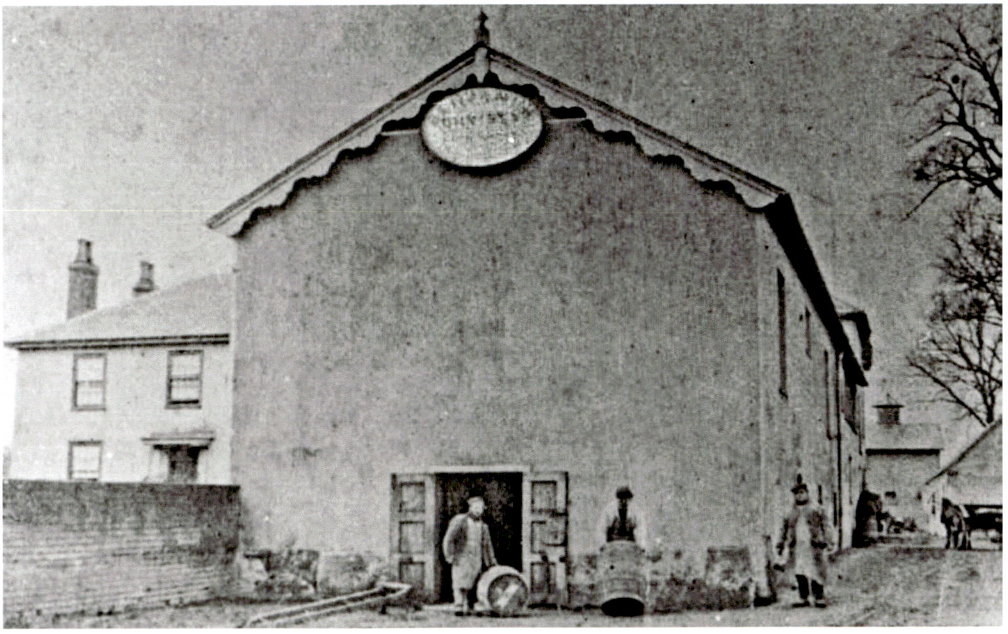

Above: Page's Brewery, Ashwell, *c.* 1890. (Courtesy of Ashwell Museum)

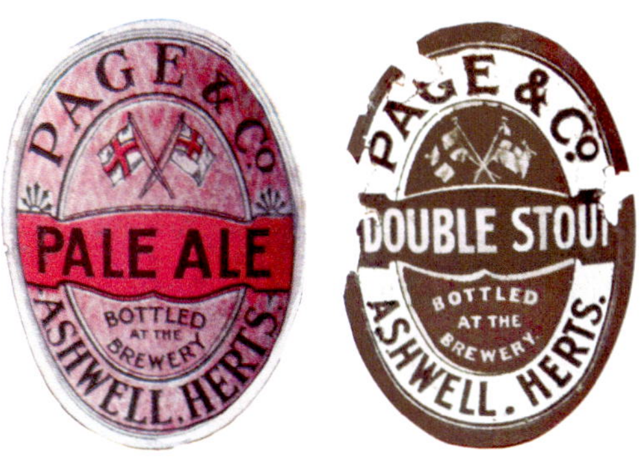

Right: Two Page & Co. labels, showing the company's 'two flags' logo. (Courtesy of Terry Knight)

BALDOCK

Ind, White Horse Street

James Ind was christened at St Ives in Huntingdonshire in 1754. In 1784 he was in Baldock and working as a brewer. By 1791 he was leasing the Barley Mow pub in Clothall, and he is listed in the *Universal British Directory* of 1794 as a farmer, maltster and brewer of Baldock, and identified as the owner of a brewery and maltings in White Horse Street, Baldock, in 1804.

James Ind's eldest son, Edward, moved to Romford, Essex, where in 1799 he went into partnership with John Grosvenor to run the Star inn and brewery there. Eleven years later, in November 1810, James Ind died, aged fifty-seven, and the brewery in Baldock passed to his only other son, James junior, born 1788. However, just nine months later, aged twenty-three, James junior was killed in an accident.

The brewery in White Horse Street and its twenty tied houses passed to Edward Ind, but he was busy building up his business in Romford (eventually to become Ind Coope), and he seems fairly quickly to have started leasing it to John Izard Pryor, owner of the brewery in Baldock High Street. In 1815 Edward sold the White Horse Street brewery to Pryor.

Pale Ale Brewery, Norton Street

John Steed was described as a brewer in November 1823 when, aged thirty-one, he took over property in Norton Street from William Oliver of Baldock, a draper. Steed, from Erith in Kent, acquired the Sun in White Horse Street in 1828, and continued adding tied houses in the 1830s. Then in June 1841 he transferred the business, valued at £560 16*s* 11*d*, to Edward Fry and George Oliver. Oliver handed the brewery back to Steed in November 1847.

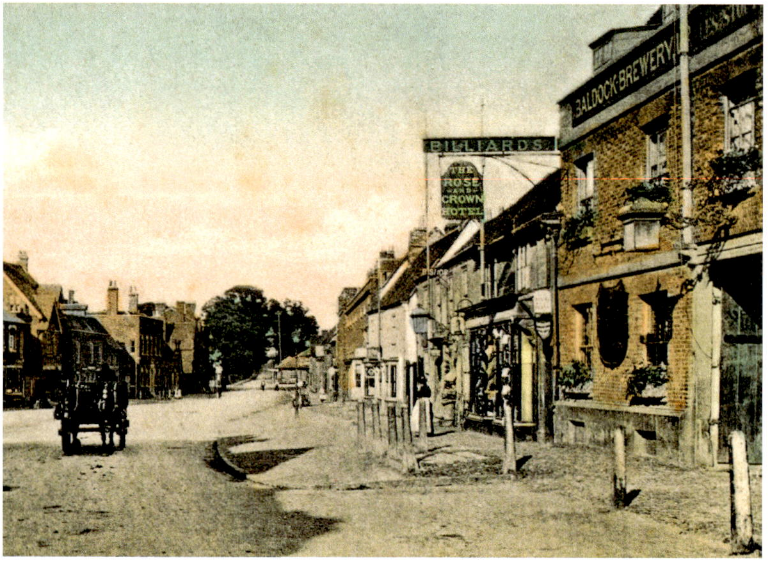

The Rose and Crown in White Horse Street, Baldock, pictured in around 1901, when it was a Baldock Brewery tied house. James Ind is said to have brewed behind the Rose and Crown in the late 1700s.

In July 1855 a worker named Cox fell into the mash tun at Steed's brewery, dying a fortnight later from 'mortification of a wound in the ancle [*sic*]'.

The business was known as the Pale Ale Brewery from at least 1870. John Steed was followed by his son, Oliver, who died in May 1888. The brewery, then making 5,000 barrels of beer a year, and its fifty-seven tied houses, was bought for £22,500 by Charles Edward Morley, acting as agent for a brewer from Burton-upon-Trent, William Pickering.

The brewery traded as Morley and Co. but seems to have struggled, and in December 1894 forty-two pubs and beerhouses were disposed of at the London Auction Mart, realising more than £31,500. In March 1898 it became the Baldock Brewery Co. Ltd, with twenty-two tied pubs and beerhouses.

In July 1904 a receiver was appointed. The same month Wells and Winch of Biggleswade bought the Pale Ale Brewery and its tied houses for £13,000. The brewery was demolished in 1969.

Simpson's

Robert Thurgood, a member of an old North Hertfordshire malting family, is first mentioned as a brewer in 1743. By the time he died in January 1775, aged sixty-one, Robert's possessions included the brewery on the east side of Baldock High Street and a dozen public houses.

His daughter, Sarah, had married another brewer, Thomas Clutterbuck, whose brewery was in Stanmore, Middlesex. Robert Thurgood left his brewery and tied houses to Sarah and her son, Robert Clutterbuck, and his descendants. But Sarah's son was only three years old. Sarah leased out the Baldock brewery business to John Pryor, a Quaker and a member of another local malting family.

The Clutterbucks eventually decided to sell the brewery in Baldock and eleven inns and pubs in 1799 to John Pryor's eldest son, John Izard Pryor, for £8,842. One source says the Georgian brewery buildings were built in 1780 and extended in 1811.

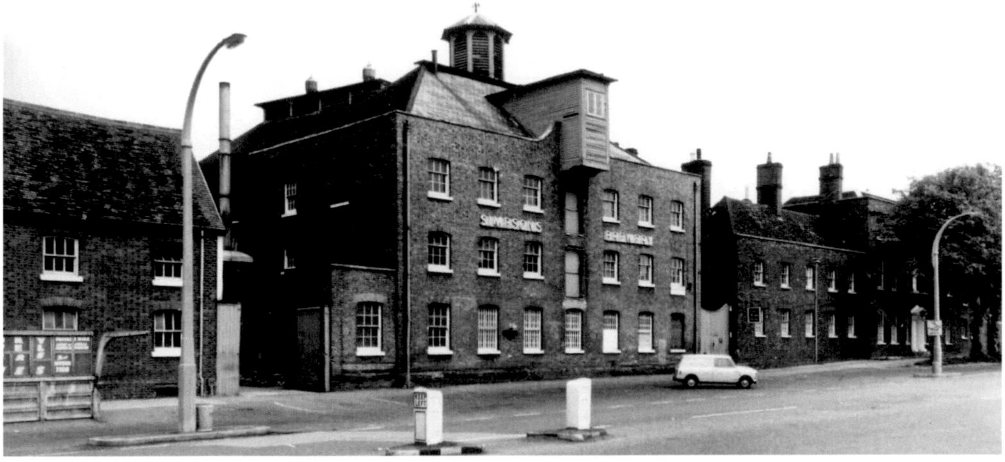

The beautiful Georgian building of Simpson's Brewery, built in around 1780, is seen just before its closure in 1965 and then demolition in 1968.

Two of John's brothers, Thomas Marlborough Pryor and Robert Pryor, had been leasing another concern, Proctor's Brewery in Shoreditch, East London (formerly Harwood's). When their lease on Proctor's Brewery ran out, in 1816, the two Pryor brothers brought their trade into the big Quaker brewing partnership of Truman, Hanbury and Buxton in Brick Lane, Shoreditch.

In 1826, aged fifty-two, John Izard Pryor retired from the brewery. His eldest son, John junior, born in 1800, died in 1853. The Pryor family put the brewery in Baldock High Street up for sale, and it and its now 122 tied houses were bought for just under £82,000 by the Simpson brothers, Joseph and Thomas George, aged twenty-nine and twenty-two, from Stamford, relatives of John Phillips at the rival Royston Brewery.

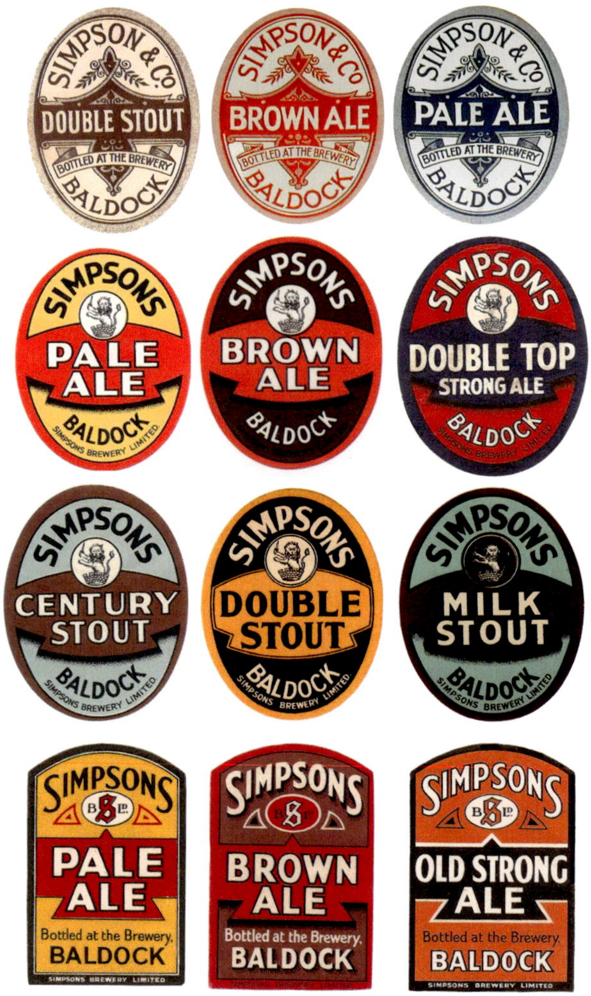

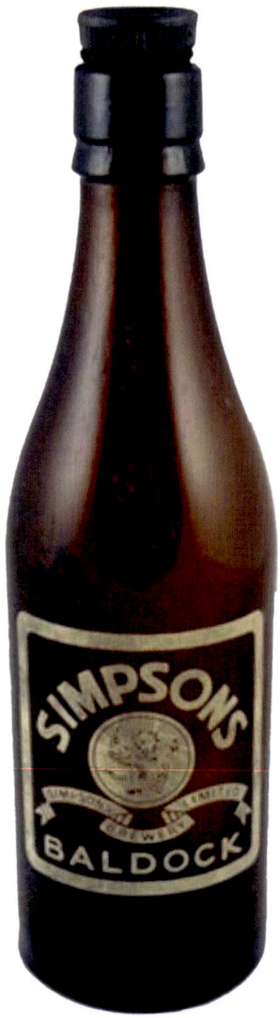

Above left: A selection of Simpson's labels. The Century Stout appeared in 1953 to mark 100 years since the Simpson brothers acquired the brewery.

Above right: A fired-on label from Simpson's Brewery.

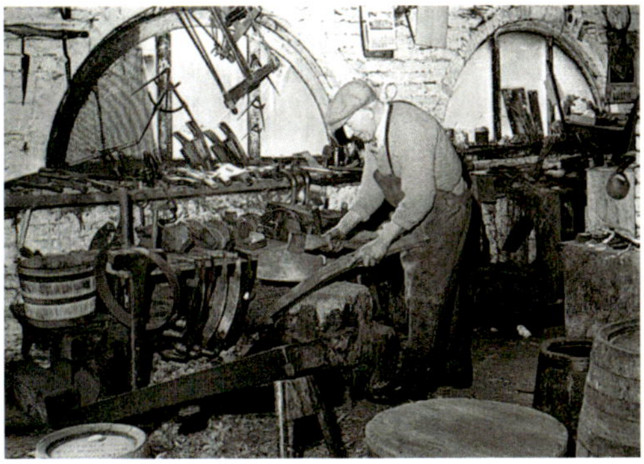

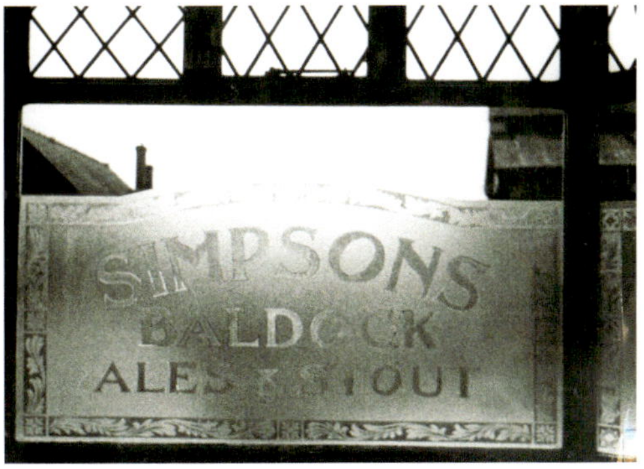

Above left: Simpson's cooper, F. G. Morris, trims a cask stave in the brewery cooperage. (Courtesy of Ego Publishers)

Above right: A window at the Mother Redcap pub, Latimer Road, Luton (photograph reversed). The original pub, named the Red Cap, was built by Simpson's in 1855. It was demolished in 1935 and replaced by a mock-Tudor pub with engraved windows advertising the brewery. This window was smashed by a drunk with a scaffold pole, *c*. 1980. The pub itself closed in 2018.

In the thirty-five years between 1855 and 1889 the Simpsons added forty-eight tied houses to the estate. But they owned no pubs in Royston, and their cousins at the Phillips Brewery there owned no pubs in Baldock.

Thomas George Simpson died in December 1889, aged fifty-eight, in London. A few years later, in October 1893, Joseph Simpson died, aged sixty-nine, leaving almost all his property to his son, Evelyn.

In June 1909 Evelyn Simpson changed his name by royal licence to Evelyn Shaw Hellier, to inherit from his maternal uncle, Colonel Thomas Shaw Hellier, the Wombourne Wodehouse estate in Staffordshire. In the First World War, Evelyn Shaw Hellier's son, Arthur, was a lieutenant with the 7th Battalion, South Staffs Regiment. In August 1915 Arthur was killed in action in the Dardanelles, aged twenty-nine. Thus, when his father died in July 1922, the brewery passed to Arthur's eldest sister, Evelyn Mary Penelope Shaw Hellier, who was then thirty-two.

Francis Tom Simpson, Thomas George's son, died in November 1929, and Miss Shaw Hellier became the sole partner. By the 1950s she was in her sixties. Negotiations began in 1953 with the Bury St Edmunds brewer Greene King, and on 1 January 1954 Greene King officially took over the Baldock Brewery and its pubs, for £525,000.

Then in 1961 Greene King merged with Wells & Winch of Biggleswade. The move doomed the Baldock Brewery, which closed in February 1965. The brewery was demolished in 1968, despite protests.

White Horse Street Brewery

William Noy owned a brewery and maltings alongside the building that was later the Victoria pub in White Horse Street before 1736. By the 1790s the brewery was in the hands of William Penn. Penn was still brewing in White Horse Street right up to 1839.

Frederick Oliver was a brewer in White Horse Street in 1841. By 1850 the brewery in White Horse Street, and an attached maltings, were being run by George Oliver. In 1851, Oliver, then aged thirty, was employing seven men at the brewery. He was also leasing the Cricketers, Ickleford, and, by 1852, the Orange Tree, Hitchin.

In 1862 James Edward Oliver and George Oliver were recorded as running the business. However, in February 1867 George Oliver, 'brewer and farmer' of Baldock, appeared before the Bankruptcy Court in London admitting debts of between £2,000 and £3,000, which he blamed on 'insufficiency of capital, bad debts and the high price of malt'.

By 1868 Alfred Cleaver, born in London in around 1845, had joined the partnership. Oliver left the following year, and in 1871 Cleaver was employing eight men at the brewery, which had twenty-six tied houses.

In July 1872 'The Hertfordshire brewery, Baldock' was converted into a limited company, with Cleaver described as the previous owner. The *Brewers' Guardian* wrote that 'an agreement between Alfred Cleaver and Burton Osborne Hilliard regulates the purchase'.

However, within eleven months the company was in liquidation.

By 1874 the brewery, with its eight-quarter plant, was being leased to Josiah William Parker. Parker & Co. were running the brewery in 1879, but in September the next year Parker announced he had given up his share of the brewery in Baldock to his 'late partner'. It looks as if Parker was followed at the brewery by William John Anstice Banister, whose business at the White Horse brewery, Baldock, was liquidated in September 1880. The brewery was put up for sale at auction in September 1881 but failed to find a buyer.

BARKWAY

William Gapes from Essex was a retail brewer in Barkway in the 1859 *Kelly's Directory*, probably at the same premises he occupied for the 1861 census, 71 High Street, now known as Brewery House.

Gapes had stopped brewing by 1862, his operation now in the hands of John Johnson Balding from Ely, son of the village doctor. The Ordnance Survey map of 1896 shows the brewery on the east side of the High Street. In 1905 the Barkway brewery had three tied houses in Hertfordshire.

The Barkway brewery was still in operation in 1908 but had vanished by 1910.

BARNET

Old Barnet Brewery Company, Wood Street

The Old Barnet Brewery Company in Wood Street claimed on its bottle labels to have been founded in 1694. In June 1805, however, the premises were described as 'formerly a brewhouse but now divided into two tenements'.

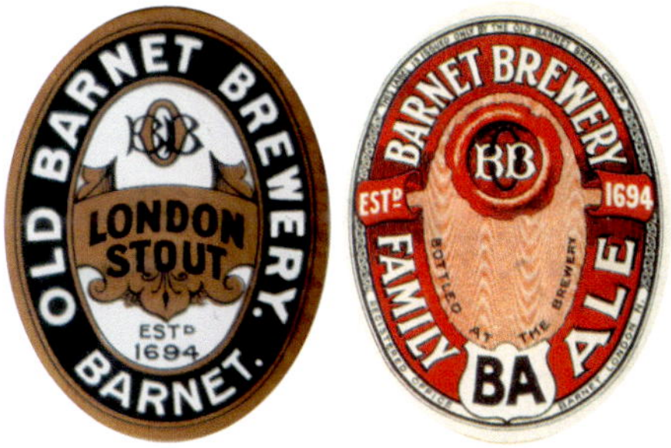

Above: Two labels from the Old Barnet Brewery. The one on the left was probably issued by Mark Buttenshaw between 1887 and 1904, the one on the right by the Leary family between 1905 and 1909.

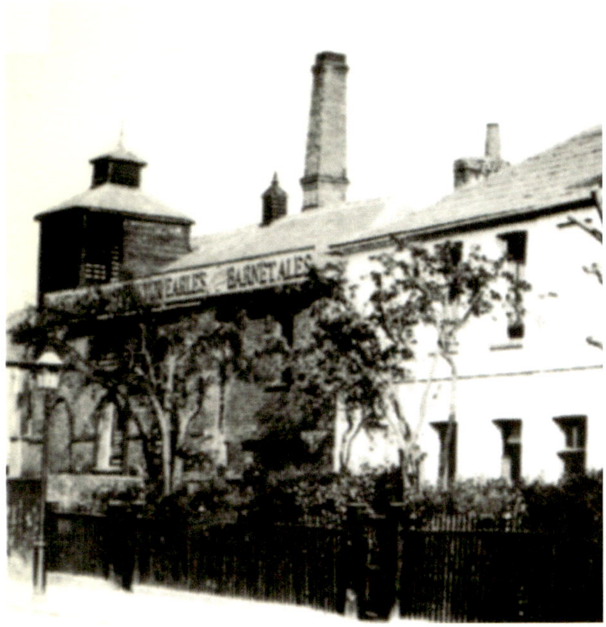

Right: The Old Barnet Brewery pictured between the mid-1870s and 1887, when it was run by Henry Earle.

By 1827 the premises were occupied by John Buckthorp, who had turned them back into a brewery. Buckthorp was still brewing in Wood Street in 1866, when he was sixty-seven. Soon after the brewery closed down. It was reopened in October 1869 by Robert Sinclair, landlord of the nearby Sebright Arms. By 1874 Sinclair had been replaced at the brewery by Henry Earle, a former solicitor, who was running it as Earle & Co.

In 1887 the brewery passed into the hands of Mark Noble Buttenshaw, a solicitor of Blackheath Park, London, who set up the Barnet Brewery Company Ltd.

A receiver was appointed in 1904 over unpaid interest, and in 1905 the business was in the hands of the Leary family. The next year they launched the Old Barnet Brewery Company Ltd to run the Wood Street brewery. Three years later, in March 1909, the company went into voluntary liquidation.

The Phoenix Brewery, High Street

William Francis Gregory appears as a brewer in Barnet High Street in a directory of 1866. Tragically, he died in July 1867, two days after slipping at the brewery and falling into a vat of 'scalding' liquor.

His widow, Mary Caroline, carried on running the brewery. Her last advertisement in the *Barnet Press* appeared on 25 September 1869, and a month later the first ad appeared for 'Nicholas Frost, late M. C. Gregory, The Phoenix brewery, High Street, Barnet'.

By 1872 Frost's son, Charles, was managing the business. The Phoenix Brewery was turned into a stores and off-licence for J. Harris Browne's brewery at Hadley Green in July 1887.

BERKHAMSTED

Swan Brewery, High Street

The Swan Inn dates from at least 1607 and was doubtless brewing its own beer from the start. James Foster was a brewer and maltster at the inn in 1839. By 1851 the inn and its brewery were being run by his widow, Hannah. Soon after this, Hannah remarried, to John Edward Lane. In 1854 Lane was advertising himself at the Swan brewery as a brewer and maltster: 'private families supplied with ale and table beer in any quantities'.

Lane started acquiring tied houses, and by the 1890s the brewery owned getting on for a dozen outlets. He handed the brewery over to two of his stepsons, Daniel and Henry Foster, in 1870, when he was sixty-two.

Henry Foster sold the brewery and its two maltings, nine public houses and three off-licences on 17 November 1897 to the Chesham Brewery Ltd for £24,000. Brewing continued at the Swan brewery until at least 1910, however.

Water Lane Breweries

There were two breweries in Water Lane, Berkhamsted, on the north side of the High Street in the first half of the nineteenth century, one on the east, the other opposite on the west. The one on the east side was evidently being run in the late eighteenth century by William Billington, who was described as a brewer when he died in 1798.

His daughter, Elizabeth, who was born in 1775, was the owner of the brewery property in 1811, when she married John Manship Mills. He took over the brewery, but early in 1841 Mills was declared bankrupt. Almost four years later, in December 1844, the 'old-established' brewery 'in the occupation of Mr J. M. Mills' was put up for auction. No more seems to be heard of brewing on the site.

The brewery on the west side of Water Lane was run, and perhaps founded, by William Tomlin, who is listed as a brewer in Berkhamsted in 1824. Tomlin ran the brewery, according to one source, until 1844. He was then followed by John Newman Frost, born in around 1803 in Westbourne, Sussex.

Frost appears in the 1852–3 electoral register for Berkhamsted, but very soon after the brewery was being leased by Alfred and Charles Frederick Healey, part of a family of brewery owners from Watford.

Alfred Healey had lost £20,000 by 1868 in bad investments and was forced to dispose of the lease of the brewery and its tied houses. The buyers were T. S. Locke and Edward Howard Markes Smith. The firm became a limited company in January 1897, when it owned forty-two pubs and beerhouses.

At Michaelmas 1898 Locke and Smith acquired the nine tied houses belonging to John Brown's brewery in Tring. On Christmas Day 1905 the company began leasing the twenty-four pubs and beerhouses previously owned by James Batchelar's Dagnall brewery.

By 1911 Locke and Smith were supplying a total of seventy-seven tied houses. However, the firm owed its banks more than £24,000. The High Court ordered

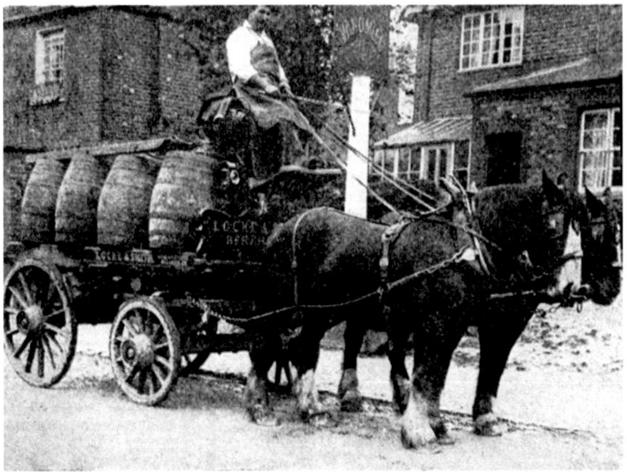

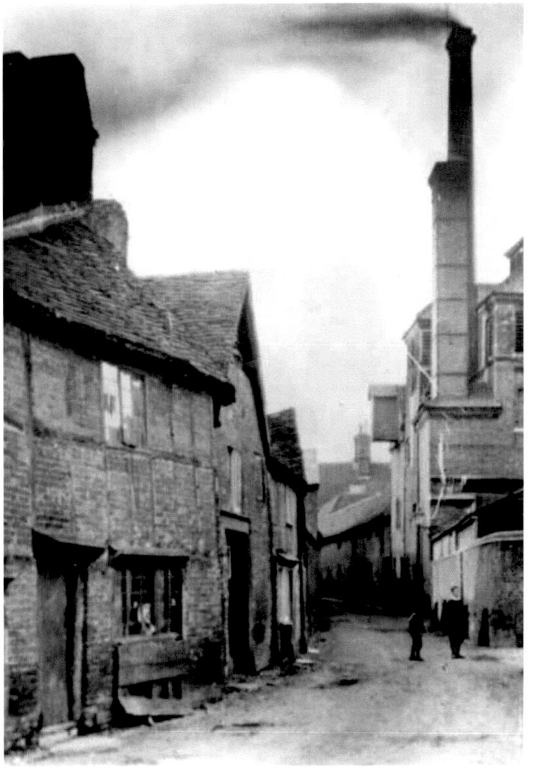

Above: Lock and Smith drayman James Miles outside a pub called the Windmill, his dray laden with wooden barrels.

Right: Locke and Smith's Brewery in Water Lane, Berkhamsted. (Courtesy of the Brewery History Society)

that the brewery and its pubs should be put up for auction. At the Auction Mart in London in June 1913 the brewery and forty-two tied houses went for £37,600 to Benskin's of Watford.

BISHOP'S STORTFORD

Fox Brewery, Dunmow Road

Joseph Robert Heath started the Fox Brewery in Dunmow Road, Stortford, shortly before 1886. In September 1890 Heath put the two-quarter brewery and an adjacent beerhouse up for sale. The purchaser, for £660, was Joseph Flinn, a maltster originally from Derby who had bought a malting operation in Stortford in around 1879.

Flinn also acquired a brewery in William Street, Herne Bay, Kent, in the mid-1890s, which was running as Flinn and Company by 1895. However, he died in June that year, aged sixty-seven. By 1898 the Fox Brewery was Flinn and Sons, a partnership of Frank Flynn and Thomas Coates Chamley.

In April 1902 Flinn and Chamley sold the Fox Brewery to William Radin Stanton of Cambridge for £2,300. Flinn and Sons controlled six tied houses in Hertfordshire, plus at least one in Essex, at Saffron Walden. Stanton was only at the brewery for four months, handing the business over in August 1902 to Marten Alfred Bailey, born in

1877, and his older brother, Harold Barber Bailey, born in 1875, sons of Frederick Bailey, who owned the Star Brewery in Cambridge until 1889.

Harold Bailey had become a partner in Bailey and Tebbutt of the Panton Brewery in Cambridge in 1897, leaving Marten at the Fox Brewery. In 1915 Bailey's was bought by Benskin's.

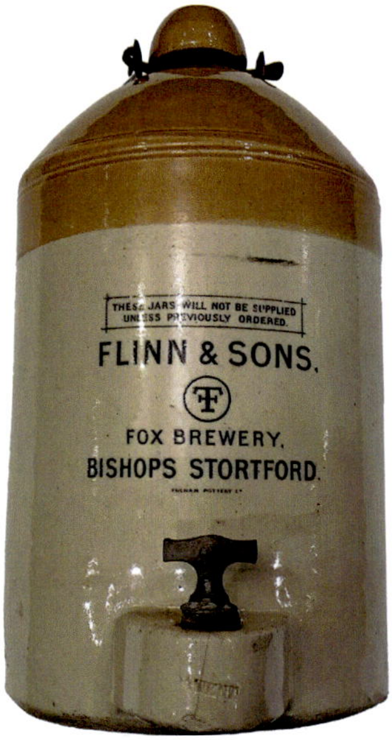

Left: A gallon beer jar, *c*. 1900. Customers would order the jars from the brewery and draymen would set out with sixty one-gallon jars on their drays to deliver to private customers. (Courtesy of Kevan Witt)

Below: Bailey Brothers' new brewery, designed by the well-known brewery architect William Bradford, as depicted in the *Brewers' Journal* in March 1906.

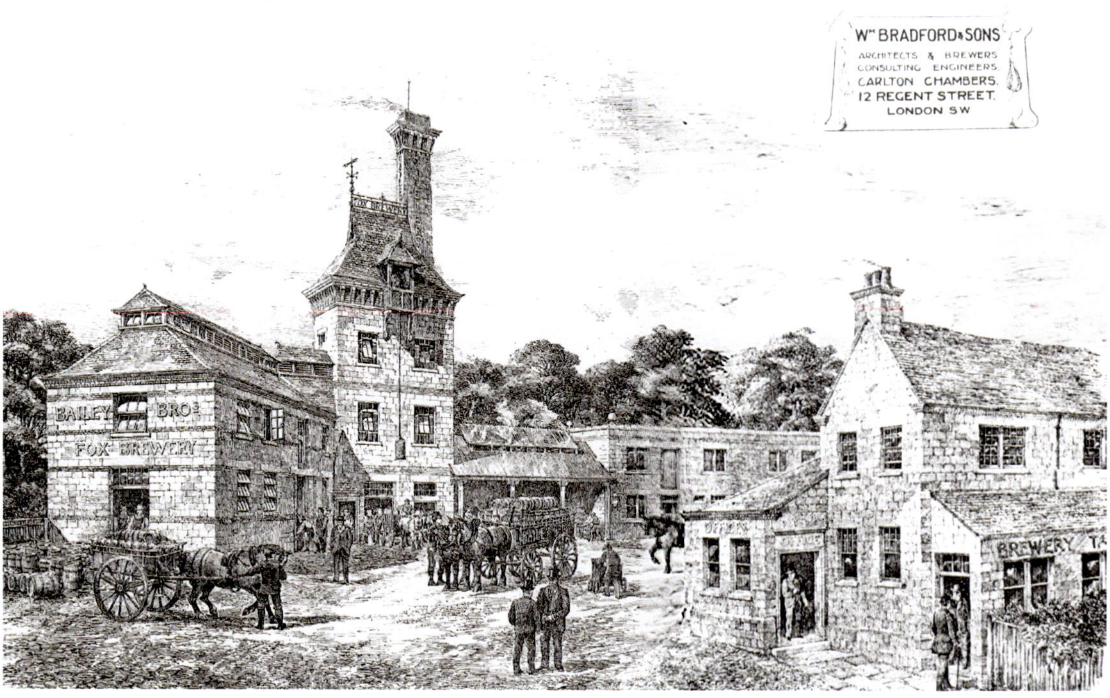

Hawkes & Co., Water Lane

William Woodham and Thomas Bird, two Bishop's Stortford maltsters, and William Hawkes, 'gentleman', built a brewery in Water Lane in around 1780.

By 1795 the Water Lane brewery owned at least six pubs in Stortford. William Hawkes was described as 'brewer and banker' when he died in February 1794, aged thirty-eight, leaving his widow, Elizabeth, to take his place in the partnership. By 1809 the partnership had been joined by Frederick John Nash, then thirty-two, from Cambridgeshire. Two years later, when Thomas Bird died, the partners included William Johnstone and Elizabeth's son, William Robert Hawkes, then aged twenty-six.

In 1836 one partner, William Johnstone II, set up his son, William III, aged twenty-six, in another concern, the Horne Lane brewery, Bedford.

William Robert Hawkes died in 1859, aged seventy-four, his place taken by his son, William Robert junior. In around 1875 the partners decided to sell up to a trio of businessmen headed by James Wigan of Mortlake, Surrey, a former partner with Charles Phillips in the Mortlake Brewery (later the Stag Brewery).

Wigan and his partners took over the running of the Water Lane brewery in May 1876, paying £100,000 and keeping the name of Hawkes and Co. Over in Braintree, Essex, another William Hawkes had bought the Black Lion, in 1869, and by 1874 he was running a small brewery business there. In 1890 the Wigan partnership ended any possible confusion by buying the Essex William Hawkes's Black Lion Brewery, together with its three other pubs in Braintree, for £1,750.

In April 1898 Hawkes bought another small Essex brewery, Smith's of Great Bardfield, with its four tied houses, for £6,600. A few months later, however, in August 1898, Hawkes was taken over itself, by Benskin's Watford Brewery, which paid the Wigans £263,000 for the concern, including some 156 tied houses.

The Water Lane brewery stayed open, and in 1916 began supplying the seventy-one or so pubs formerly owned by Taylor's of Saffron Walden, after Taylor's purchaser,

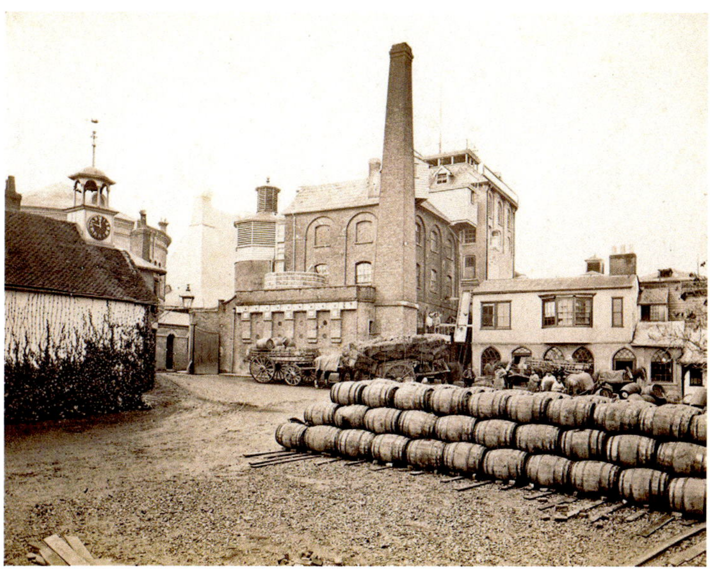

Hawkes' Brewery, *c.* 1880. The brewery clock is still there, although the building it stands on was converted into a Waitrose supermarket in 1994.

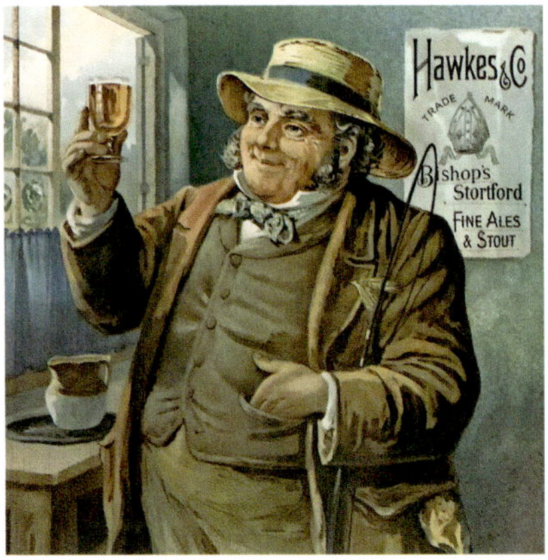

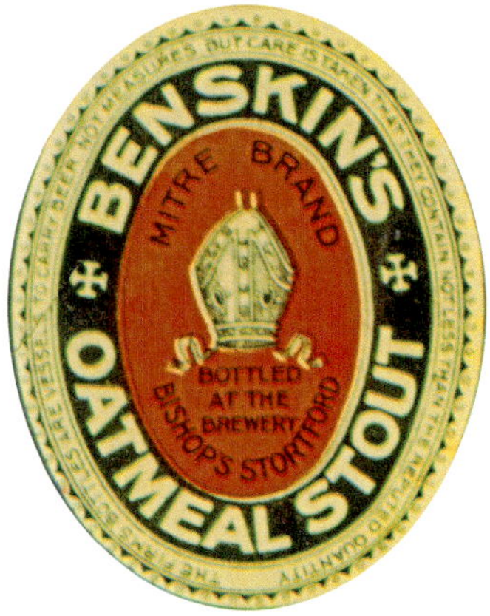

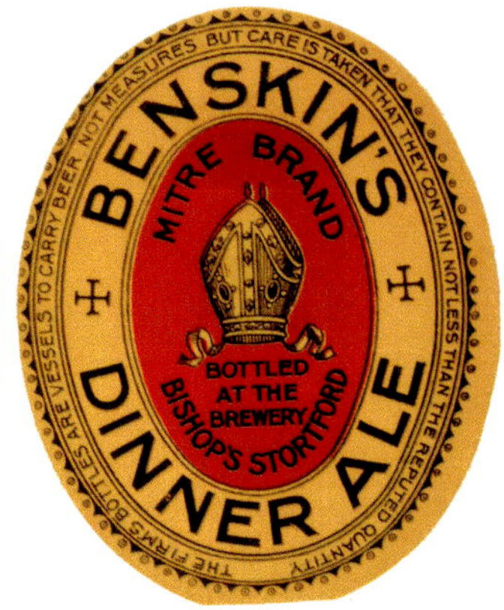

Above left: A poster for Hawkes & Co. from the nineteenth century showing a contented farmer – note the ear of barley in his lapel. (Courtesy of Kevan Witt)

Above right: Two labels issued by Benskin's after it took over the Bishop's Stortford Brewery in 1898, using the old Hawkins 'mitre' trademark.

Watney Combe Reid of the Stag Brewery in Pimlico, handed over control of them to Benskin's.

Brewing finally stopped at Water Lane in November 1921, but the site continued as a depot for Benskin's, and later Ind Coope/Allied Breweries, until January 1987.

Rye Street Brewery

Edmund Wyman, aged fifty-six, born in Little Hormead, was listed as a brewer in Rye Street in the 1851 census. Wyman died at his brewery in June 1863, aged sixty-nine. He was followed in the business by his son, Edmund junior, born in 1842 in Hoddesdon.

In February 1865 Edmund junior was arrested for debt, declared bankrupt and briefly jailed. He handed over the running of the brewery to his brother-in-law, Waller Maber King, a tobacconist and newsagent in North Street, Bishop's Stortford. King was prosecuted for running the brewery without a licence in September 1865, but the case was dismissed.

The brewery, mashing three quarters of malt at a time, was still running after Wyman was released, but in November Wyman advertised the brewery and the lease of its tap and stores 'to be disposed of'. No more seems to be known of it.

BOREHAMWOOD

Star Brewery

John Cockburn, forty-eight, from Reigate, Surrey, was brewing in Theobald Street, Borehamwood, by 1870. The last time Cockburn's brewery is listed in *Kelly's Directory* is 1886. By 1889 a former employee of Cockburn, Alfred Parkins, forty-five, had set up a brewery in two converted labourer's cottages across the road, next to the Wellington pub.

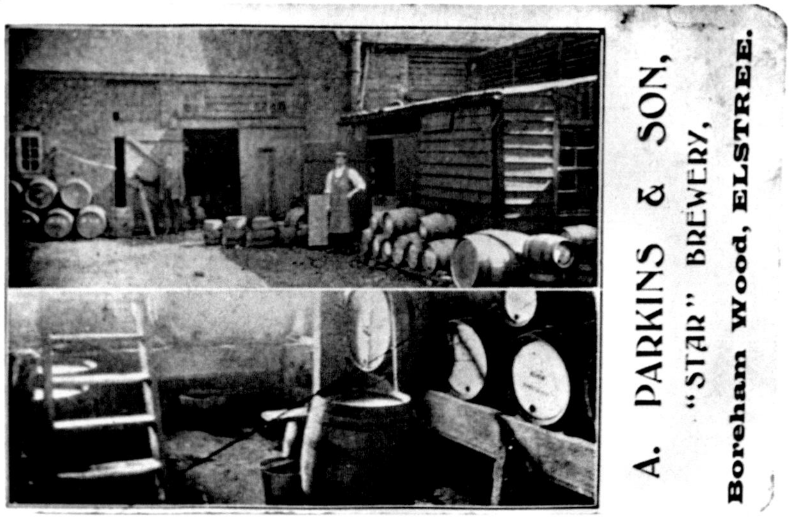

A showcard for the Star Brewery at Borehamwood. Could that be Alfred Parkins by the door and his unlucky son, Ted, in the brewer's apron?

Alfred's brewery – known alternatively as Parkin's Sons or Parkinson Brothers or Parkins & Son – employed four men, with two small pony drays to deliver the beer. In November 1897 Alfred's son, Ted, then in his mid-twenties, was killed when the pony hauling the dray out to customers at Shenley suddenly bolted. The dray rolled up the bank at the side of the road and a cask of beer dropped down onto Ted, fatally crushing him.

By 1905 the operation was known as the Star Brewery. Alfred died on 3 January 1909, aged sixty-five. The brewery was sold to a man called H. Sanders, and in May 1909 a new company was registered called the Elstree Star Brewery Ltd. However, on 5 October 1911 a receiver was appointed to wind up the company.

CHESHUNT

Lion Brewery

James Griffiths was running the Lion Brewery, Cheshunt Street, from at least 1861. He died after 1866, and his widow, Elizabeth, retired in 1869. Her son-in-law, William Leighton, announced he would carry on the business. However, by 1871 he was only a 'beer dealer'.

Edward Sharpe

Edward Sharpe began brewing beer in Cheshunt Street in around 1850, calling himself the Cheshunt Brewery. By 1855 Sharpe had been declared an insolvent debtor.

Spurway

The Spurway family had started another Cheshunt brewery by 1866. William George Spurway was at the brewery in 1874, which vanishes after that.

FURNEUX PELHAM

Rayment's

In 1677 Furneux Pelham Hall was acquired by the Calvert family, later owners of the Hour Glass Brewery, one of the biggest porter breweries in London. Around the time of the Napoleonic Wars, the Rayment family began leasing the hall. At some point in the 1850s the then tenant, William Rayment, moved into commercial brewing. In 1858, when the Calverts put the hall up for sale, prospective buyers were told that 'attached to the premises is a compact brewery, with excellent stores, capable of carrying on a trade of 20 to 25 quarters per week', that is about sixty barrels or so.

By 1859 William was organising the erection of a purpose-built brewery by the crossroads at Barleycroft End, in the eastern part of the village, with an artesian well practically under the brewhouse.

The brewery opened in 1860. However, William had suffered from tuberculosis for many years. He died on 10 December 1861 at Furneux Pelham Hall, aged forty-eight.

By 1871 the brewery was under the control of a manager, Charles Scruby, who lived at Mill House in the village. The Brewery Tap beerhouse had opened up on the opposite side of the road to the brewery premises.

Then, in three years from 1873, three of William Rayment's four sons died in succession. But William's daughter, Fanny, still lived in the village. By 1878 she and her husband, John Pottrill, were living at Pelham Lodge Farm, which was in the same ownership as the brewery. Pottrill evidently ran the brewery with Scruby as his manager. In the 1880s the Pelham brewery had an estate of close on twenty tied houses in Hertforshire and Essex.

In February 1888 Frank Rayment, the last of William's sons, died aged thirty-four. In December the brewery and its pubs were put up for sale at the Auction Mart in

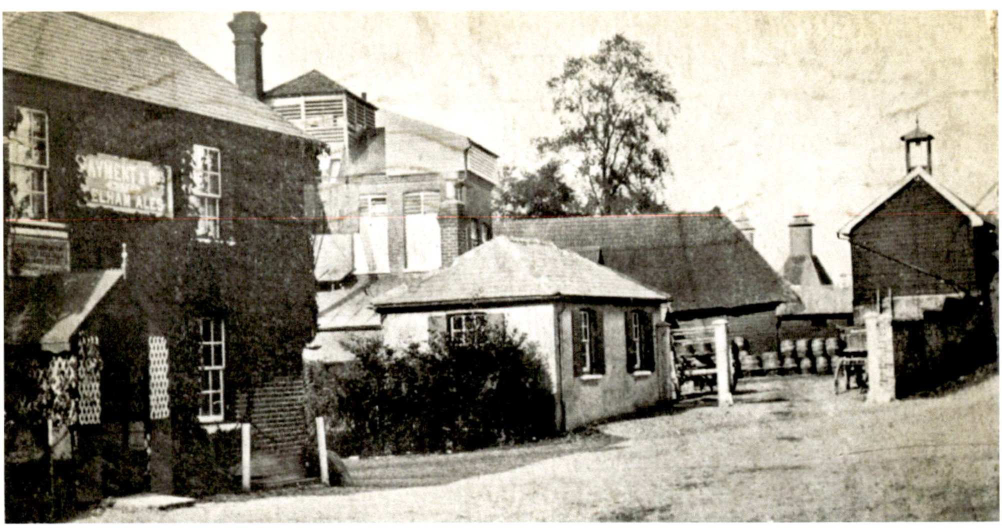

Rayment's Brewery, *c.* 1920. The thatched brewery store and malt kilns are visible behind and on the left is the Brewery Tap.

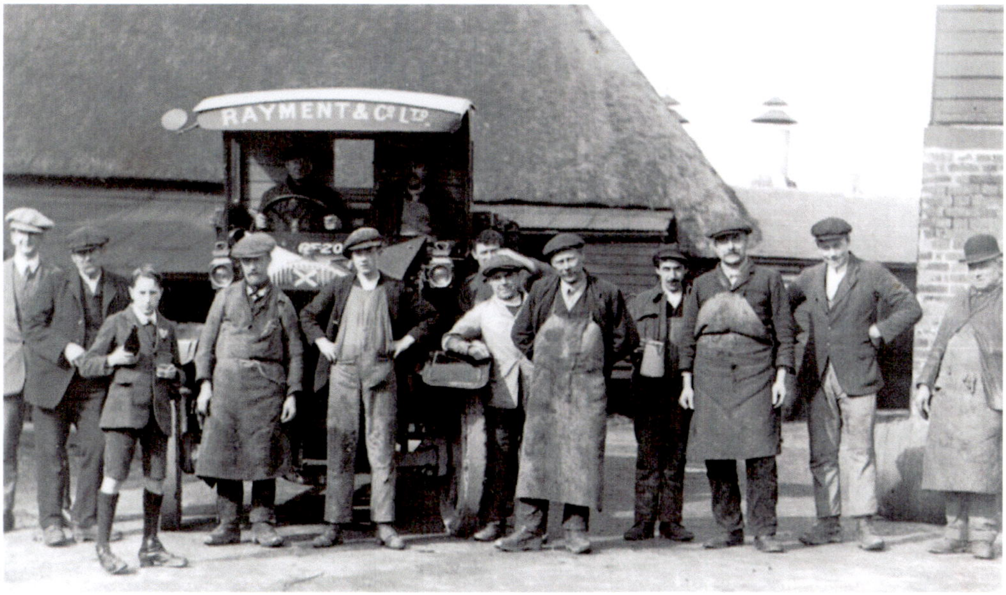

Above: Rayment's staff and workers pictured in the brewery yard, probably in the early 1920s. Is that boy old enough to drink that beer?

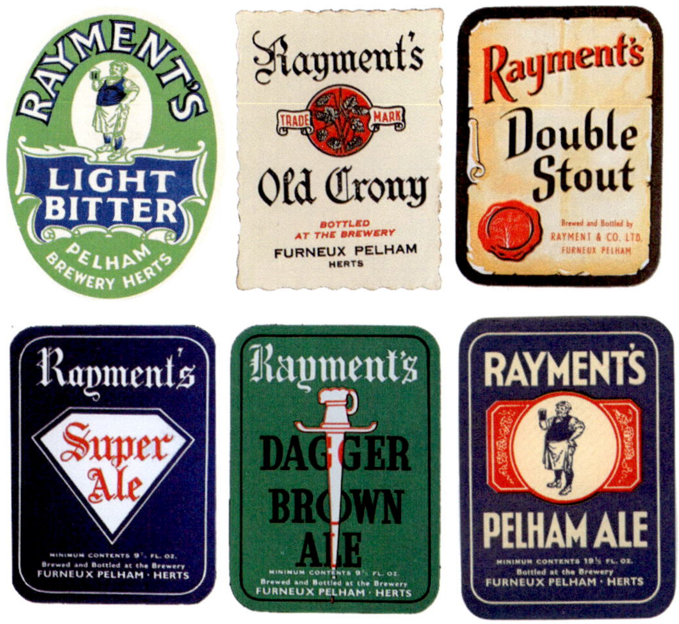

Right: A selection of Rayment's labels, including two showing the 'little brewer' figure that became its trademark, though the same image was used by two other brewers, one in the West Country, the other in Ireland.

London. The buyers were Edward Lake and James King, who paid around £26,000 for the brewery and its one leasehold and eighteen freehold pubs and beerhouses, rolling stock and so on.

Lake and King were the joint managing directors of Greene King Ltd of Bury St Edmunds, which had been formed by the amalgamation of two neighbouring rival brewers in the Suffolk town only the year before. They bought Rayment's as a private concern completely separate from the Bury St Edmunds operation, however.

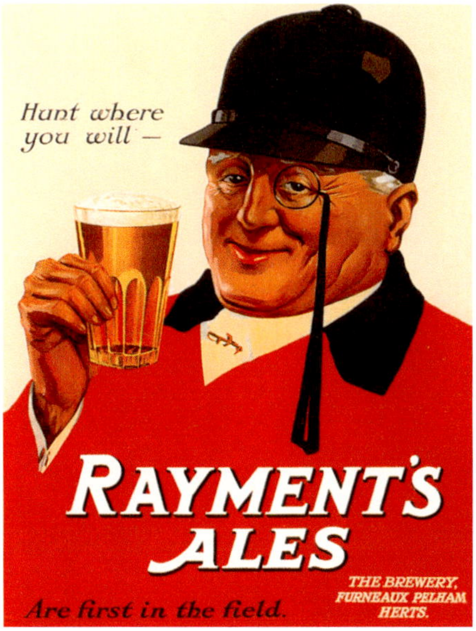

Rayment's used calendars and posters bearing the red-jacketed huntsman figure designed by the printers Joseph Causton that was used by both Tetley's in northern England and Eldridge Pope in the west. There was just one problem with the Rayment's version – the printer spelt the name of the brewery's home village wrong …

Edward Lake's third son, Alan, twenty-five, took over as manager and brewer at Pelham in 1912. James King relinquished all links with Rayment's, and the brewery was turned into a limited company controlled by the Lakes.

Alan Lake died in 1930, at the age of forty-four. His eldest brother, Edward Lance Lake, was now chairman of Greene King, and in the autumn of 1931, despite the doubts of other Greene King directors, Rayment's and its thirty-six-odd pubs and beerhouses became a subsidiary of Greene King.

The Pelham brewery continued to run as a virtually independent Lake family enterprise. Edward Lance Lake's cousin, Vice Admiral Walter John Challoner Lake, took over the running of Rayment's in 1931, staying until 1937. His place was taken by Edward Lake senior's fourth son, another navy man, Captain Henry Neville Lake, an ex-submariner, who was in charge for thirty years.

In 1976 a second-hand fermenting vessel from the old Charles Wells Horne Lane brewery in Bedford was installed at Pelham to increase capacity, and £100,000 was spent on a new sewage plant to cope with the brewery wastes. However, in April 1987, Greene King announced that brewing was to cease at Furneux Pelham.

HADLEY

Hadley Brewery

Claims were made that the Hadley Brewery was 'established in 1700' or dated from 1754. However, no building looks to have occupied the site until 1764. Only in 1780 is there a mention of a 'new brewery', when the occupier was David Hall, who was there until at least 1806.

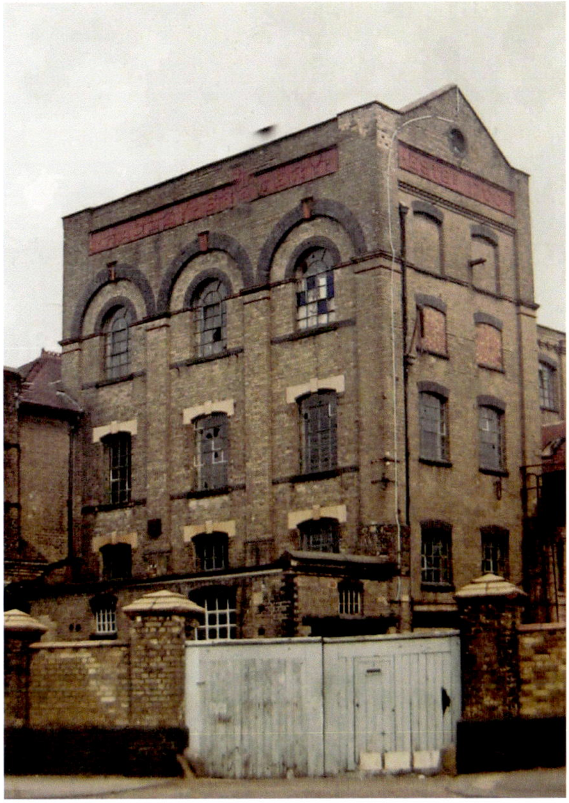

Right: The Hadley Brewery, proudly claiming to be 'Estbd 1700', shortly before its demolition in the late 1970s. (Courtesy of the Brewery History Society)

Below: A late nineteenth- or early twentieth-century showcard for J. Harris Browne's bottled beers.

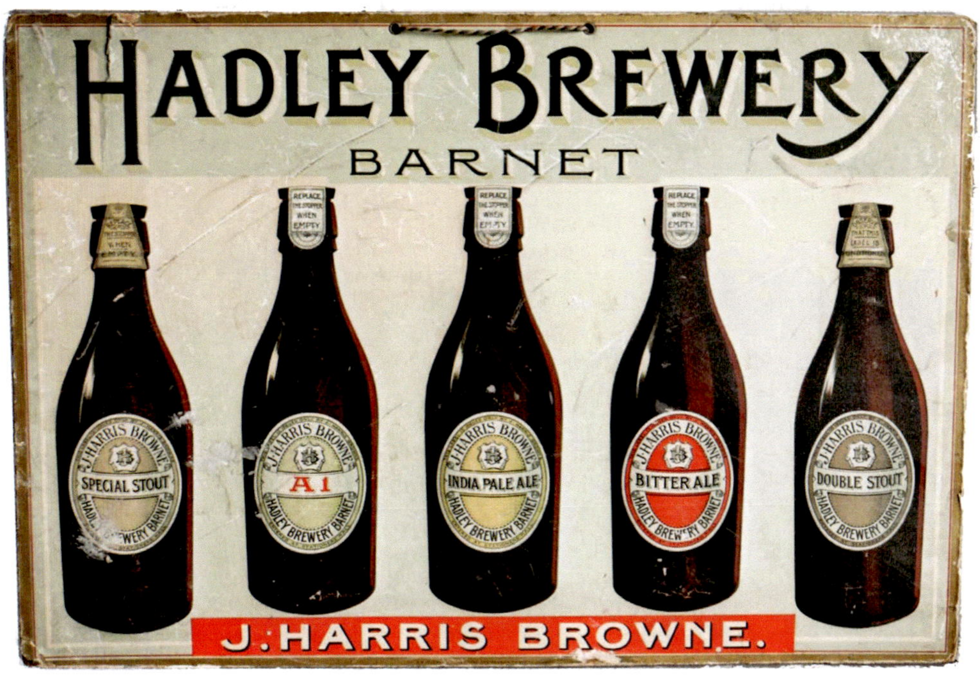

From 1809 the brewery was run by Robert Harris. He was followed in 1814 by Richard Thorp. The Thorps owned the brewery until shortly after 1846.

Addison and Co. were running the brewery in 1850, them Henry Salisbury, aged thirty-nine, from Clerkenwell, in 1851. Salisbury left around the beginning of 1861, and by March that year the brewery was in the hands of William Thomas Healey of the Watford family of brewery owners, aged twenty-seven.

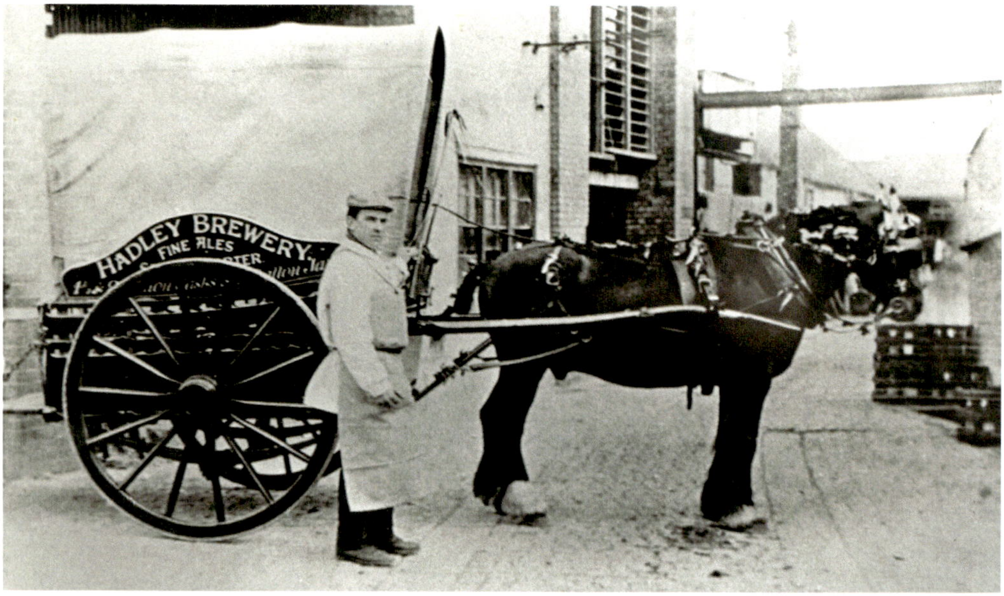

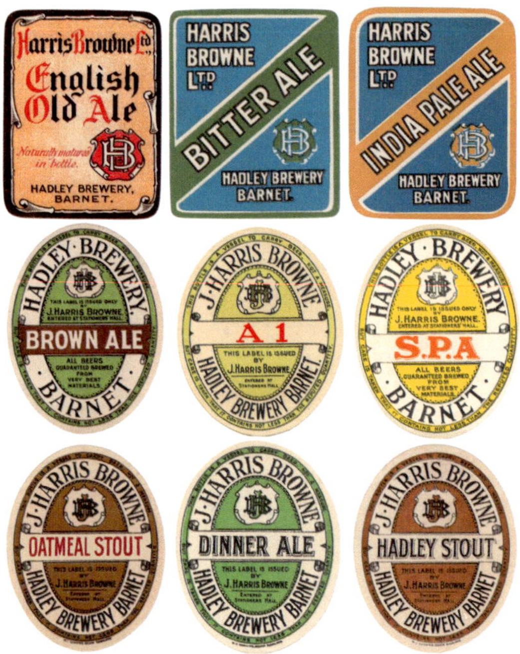

Above: A pony dray of the sort used to supply customers with one-gallon jars of beer stands in the Hadley Brewery yard, probably soon after the brewery was rebuilt in 1911.

Left: A selection of Harris Browne labels, the top row post-1930. (Courtesy of Keith Osborne)

Healey stayed two years. Between December 1863 and November 1884 Hayward Edwards was in charge. He was replaced by W. R. Dagnall & Co., whose place had been taken by January 1887 by James Harris Browne.

Harris Browne had trained as an architect, and in 1911 he had the Hadley Brewery entirely rebuilt to his own design. He died in 1932. Six years later J. Harris Brown Ltd and its four pubs were taken over by Fremlin's Brewery of Maidstone, Kent. The brewery site was still a Fremlin's depot when that company was acquired by Whitbread in 1967. It was shut in February 1969 and demolished in the late 1970s.

Two Brewers, Hadley Highstone

Joseph Buckland was at the Two Brewers, Hadley Highstone, by 1846, and brewing there in 1854.

In 1855 he was listed as 'Joseph Buckland, Two Brewers, brewer and omnibus proprietor, Highstone Brewery'. He was still brewing at the pub in 1870, and still the landlord in 1874, but by 1878 a new landlord had arrived and brewing evidently stopped.

HARPENDEN

Harpenden Brewery

John Lines, born in 1779, opened a brewery on the east side of the High Street sometime around 1806. Ten years later, aged thirty-seven, in June 1816, he died of internal injuries received after falling into an empty vessel at the brewery.

Lines's widow, Jane, aged thirty-five, then ran the brewery. She married a local surgeon, Francis House Kingston, in 1821. In November 1836, the brewery, then producing 2,000 barrels a year, and 'several' public houses and beerhouses were put up for auction.

This may have been prompted by the opening of a rival brewery three doors to the north by Francis Kingston's brother, William, that year. The buyer appears to have been James Curtis, a Quaker living in Kingston, Surrey.

Curtis died in February 1852, aged forty-one, and by 1853 the brewery had passed to George Healey, son of a Watford grocer. George died in December 1861, aged forty-three, and the reins were picked up by his widow, Elizabeth.

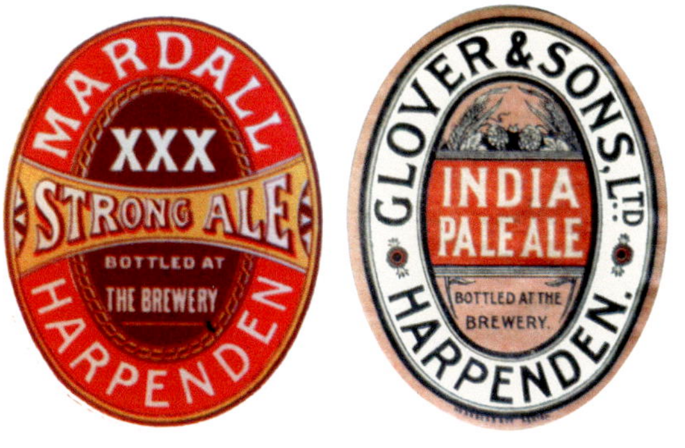

Labels from Mardall's Brewery, Harpenden, and the company that replaced it, Glover & Sons. (Courtesy of Keith Osborne)

By 1871 Elizabeth was leasing the brewery and its tied houses to Benjamin Bennett, a maltster of Dunstable.

Elizabeth died in February 1892, aged sixty-six. The Harpenden Brewery, plus eighteen pubs and beerhouses, were bought in 1893 by Henry Mardall of the rival Peacock for £16,000.

Peacock Brewery

William Kingston, a veterinary surgeon, died soon after opening the Peacock Brewery, and it was put up for sale in December 1837.

The purchaser was evidently John Isaac House of Wheathampstead. The brewery seems to have been leased to a Mr Complin in the early 1840s. He died in 1843, and John Isaac and James House were back at the brewery by 1846. John's widow, Elizabeth, then fifty-six, was still living at the brewer's house in 1871 and running the business.

Later that year the Peacock Brewery came into the hands of James Mardall, aged about forty-eight, a farmer born in Wheathampstead. The brewery eventually passed to his son, Henry.

In November 1897 the brewery, now producing 5,000 barrels of beer a year, and thirty-three pubs and beerhouses were put up for sale. The buyer, for £58,500, was Richard Alfred Glover, chairman of the Wenlock Brewery Company Ltd of Shoreditch, London, who formed Glover & Sons Ltd.

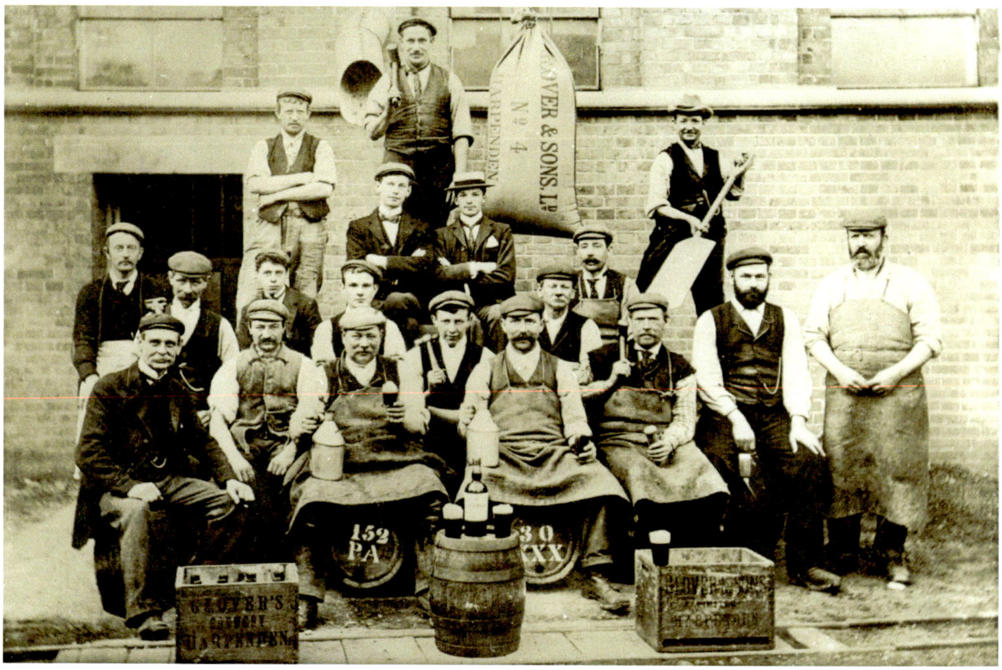

Staff and workers at Glover's Brewery, probably just before the First World War. Several are carrying the tools of their trade, including the maltster with his wooden shovel.

Glovers struck a deal in 1902 with its rival, Pryor Reid of Hatfield, to merge under the name of Hatfield and Harpenden Breweries Ltd.

With the death of his son Geoffrey Reid in the First World War, the company's chairman, Percy Charles Reid, decided to sell the breweries. In March 1919, J. W. Green's Brewery in Luton made a successful offer for fourteen of Glover's pubs and five of its beerhouses. Another seven or so Glovers outlets were auctioned off in July 1919. Finally, in January 1920, eleven ex-Glovers tied houses were acquired by Benskin's of Watford.

HATFIELD

Newtown Brewery

Started by the brothers Charles and George Bradshaw behind the White Lion, Hatfield Newtown, about 1850. Acquired by Pryor Reid in August 1888.

Park Street Brewery

The Park Street brewery started as a maltings around 1700. By 1829 William Pratchett, maltster, was brewing as well. In 1837 Francis Denyer Complin, from Hampshire, acquired the lease on the brewery. In September 1861 the brewery and its maltings were put up for auction on the instructions of Mrs Complin, 'who is retiring from business'.

By 1866 the property was being leased to James Thatcher. The lease was transferred to Arthur Sherriff in 1872, who sold out to Edward Pryor in 1878.

Pryor Reid

The Searancke family were Flemish immigrants. As early as 1589 John Searancke 'the bruar' is mentioned, apparently living at Hatfield Woodside. In 1607 John Searancke, yeoman and 'brewar', is recorded in the Hatfield manor rolls. This is presumably the same man who owned a brewery in 1614 at the place still known as Brewhouse Farm, Wildhill, Hatfield, and who died in 1617. It may well have been his son who was the Thomas Searancke running the Chequers inn at the foot of Fore Street, Hatfield, in 1635.

Four more generations of Searanckes brewed at the Chequers and built up the business, until John Searancke, who died childless in 1779. He was a JP, deputy lieutenant of the county and owner of considerable property, including 1,000 acres in Philadelphia. In 1777 his name was 'pricked' and he became High Sheriff of Hertfordshire for a year.

The Hatfield brewery property eventually passed to his nephew, Francis Carter Niccoll junior, son of a St Albans brewer, Francis Carter Niccoll. In order to inherit the brewery, however, in 1781 Francis had to seek royal permission to change his name, under the conditions of his uncle's will, to Francis Carter Searancke.

Francis was in partnership at the Hatfield brewery with John Bigg. In October 1815 Francis sold his share in the Hatfield brewery to his partner Bigg for £11,154. Bigg died four years later, bankrupt. The brewery, and twenty pubs, were bought by Joseph

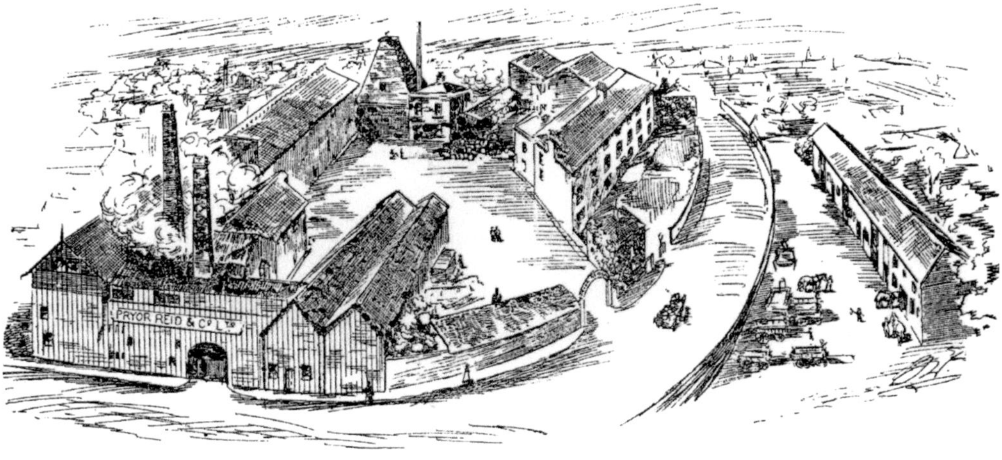

The Hatfield Brewery depicted in 1907 in the *St James's Review*, showing the maltings and malt kiln as well as steam rising from the brewhouse.

Field of Northchurch, who paid £19,500 for everything except the plant and stock. Before he died in January 1836, Field took the number of tied houses up to thirty-four.

The brewery was put up for sale by auction, and the successful bidder was James Spurrell, a brewer and hop merchant, and brother-in-law to James Watney (who soon after acquired his own brewery in Pimlico).

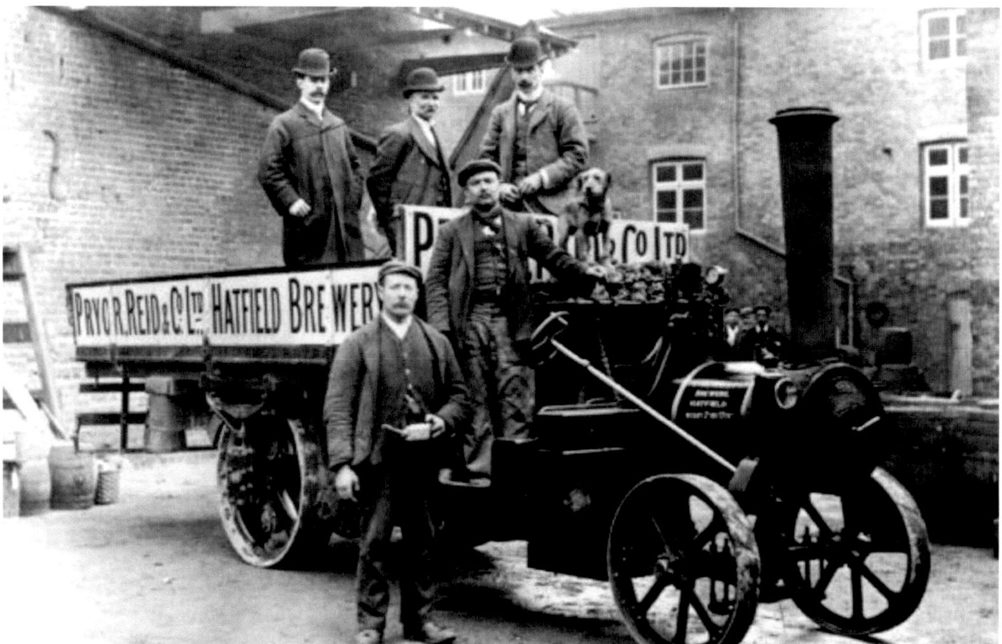

A steam dray, with very likely members of the Reid family on board. Judging by the absence of a number plate, this picture was taken before 1904.

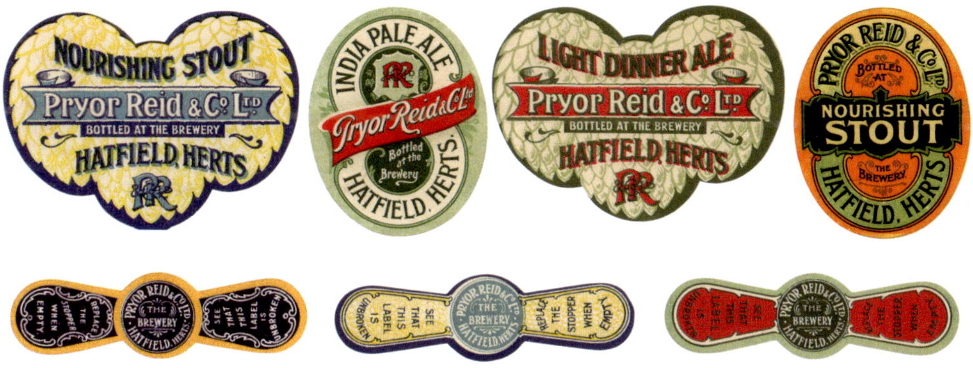

A selection of labels from Pryor Reid.

However, eight months later, in January 1837, the Hatfield brewery came on the market again. This time, it was bought by John Izard Pryor, the Baldock brewer, for £31,000 for his youngest son, Alfred.

Alfred Pryor died in 1876, aged sixty-five. His younger son, Edward Vickris Pryor, took over the running of the brewery. In 1871 he had married Ethel Justinia Reid, daughter of the late William Reid of The Node, Codicote, and the owner of the Griffin Brewery, Clerkenwell. In or shortly before 1883 Edward Pryor took his brother-in-law, Percy Charles Reid, William's youngest son, into partnership at the brewery in Hatfield.

In the autumn of 1893, Pryor Reid almost doubled in size with the takeover of Benjamin Young's Brewery in Fore Street, Hertford. Four years later, Lattimore's Brewery, Wheathampstead and its tied estate (at least three pubs) were acquired.

In spring 1902, the Hatfield company amalgamated with Glover and Sons Ltd of the Harpenden Brewery. The two breweries continued trading under their own names.

Percy Reid's only surviving son, Geoffrey Percy Nevile Reid, was killed in action during the Second Battle of Ypres, aged twenty-seven, in 1915. Business continued until the war was over, but a piecemeal sell-off of the business started in March 1919. Finally, in January 1920, the Hatfield brewery and 105 more pubs, including eleven previously belonging to Glover's, were bought by Benskin's of Watford, already the county's biggest brewers, for £190,000.

HEMEL HEMPSTEAD

Alma Brewery, Marlowes

Caleb Beckett was a brewer aged twenty-eight living in Hemel High Street in 1841, and apparently working for someone else. Not until 1861 was he definitely running his own concern, the Alma Brewery, behind the Sebright Arms.

The Becketts were still at the Sebright Arms in 1881. However, in September that year the brewery plant at the Alma Brewery was put up for auction, and the brewery vanishes.

Anchor Brewery, Bury Mill End

W. J. Liddon was listed in 1822 as a brewer in Hemel. Eleven years later John Liddon owned a brewery and maltings at Bury Mill End.

By 1845 Liddon had gone, apparently replaced by Thomas Elisha Deacon. In 1851 Samuel Deacon, twenty-nine, was a brewer at the Bury Mill End brewery. In or just before 1859 the brewery was acquired by Thomas Clement, aged twenty-six. He was there in 1862, but by 1866 he had been replaced by Alfred Wood. He, however, was bankrupt by 1871, and the brewery in the hands of Joseph Hales.

Hales, who seems to have been the first to call the business the Anchor Brewery, was there until at least 1877. He left a tied estate of some twenty-three beerhouses and just one pub, the Bell, Hemel. The following year the brewery was being run by Henry Wyman, twenty-eight, from Kettering, and Charles Addison Hall, twenty-six, from Clifton, Gloucestershire.

In 1884 the partnership ended, and Henry Wyman continued on his own. In summer 1891 he sold the Anchor Brewery and its pubs to Nash & Co. of the Chesham Brewery.

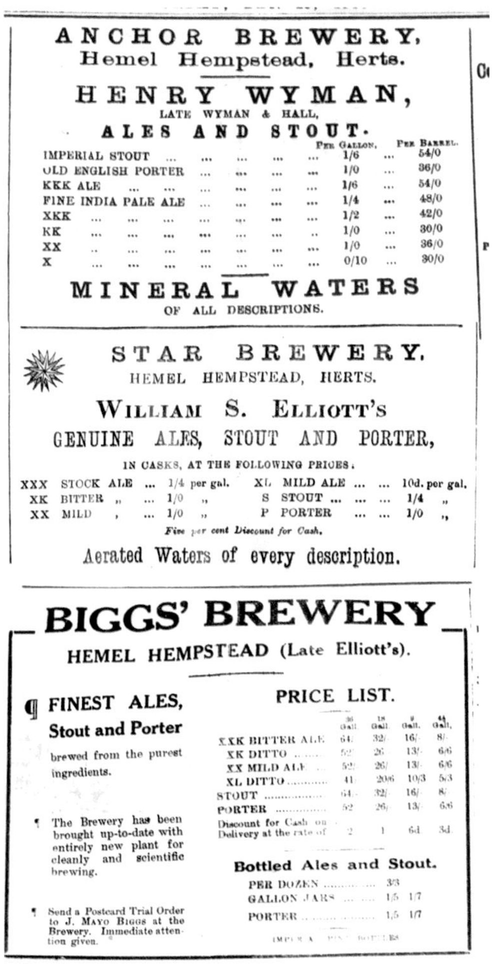

Three newspaper advertisements for Hemel breweries: Henry Wyman, the Anchor Brewery, 1885, with imperial stout and KKK ale on sale at a price that suggests an abv of around 6½ to 7½ per cent and a pub price of seven pence a quart pot, IPA at around 5½ to 6 per cent, porter, KK light bitter and XX mild ale at 4½ to 5 per cent, and a retail price of four pence a pot, and X ale at 3 to 4 per cent and three pence a pot; William Elliott, the Star Brewery 1886, with nothing stronger than 6 per cent abv; and John Mayo Biggs at the former Elliott Brewery in 1916, with war taxes already having put the cost of beer up by between 30 and 40 per cent.

Boxmoor Brewery

William Deacon appeared as a brewer at Boxmoor in 1845, and was still there in 1855. The business then went through a swift succession of owners: J. Turner and Sons around 1862, Walter W. Hamblen by August 1865 and Thomas Jeffrey junior, son of the landlord of the Rose and Crown, by 1869.

Jeffrey sold to Horatio Warren of Northampton in December 1875, but Warren appears to have left the business by the end of 1876. In 1878 Anthony Merry, thirty-three, son of Robert Merry, a Hemel Hempstead physician, was brewing at Boxmoor. He died soon after, and the brewery vanished.

Star Brewery, Bury Mill End

John F. Bontems was a brewer and coal merchant in Hemel High Street in 1851. In 1854 Bontems is recorded as a brewer and coal merchant at Bury Mill End. He was mentioned as an ale brewer in 1856, but gone by 1859.

His replacement looks to be James Elliott, aged around thirty-five, born in Aldbury. James was joined by his eldest son, William Spicer Elliott, in 1871, when William was twenty-one.

William Elliott ran the brewery until 1915, when he was sixty-five. It was sold to John Mayo Biggs, aged thirty-seven, a brewer from Dorchester, Dorset. Biggs decided to enlist in 1917, and the licence of the Star Brewery was transferred to Charles Mantle, formerly a beer retailer at the Jolly Drayman, Bury Hill. Private Biggs was killed on the Western Front at the end of March 1918. Mantle never seems to have brewed at the Star Brewery.

Two Waters Brewery

Dr Hugh Smith of Hemel Hempstead left his brewhouse in Two Waters to his daughter Lydia Holloway in 1784. The brewery was run from 1785 by Lydia's son, George. When George died in 1827, he had built up a tied estate of five pubs, which look to have been left to his illegitimate daughter, Lydia, since she married William Weller of Amersham, and all five of George's tied pubs were later Weller pubs.

The brewery at Two Waters was put up for auction in September 1827. The buyer looks to be Isaac Winter, who was listed as a brewer at Two Waters in 1832. In 1841 he was fifty-eight, a brewer living at Duck Hall, Two Waters. By December 1842, however, he was bankrupt and living in Chesham.

HERTFORD

William Baker, Hope Brewery, Old Cross

William Baker bought the Old Cross house, Hertford, and maltings alongside in 1830. By October 1831 he was being described as a retail brewer.

He was still in charge of the Old Cross brewery in 1834, but by January 1836 John Adams Cater was at the Hope Brewery. When Cater died in February 1847 the brewery

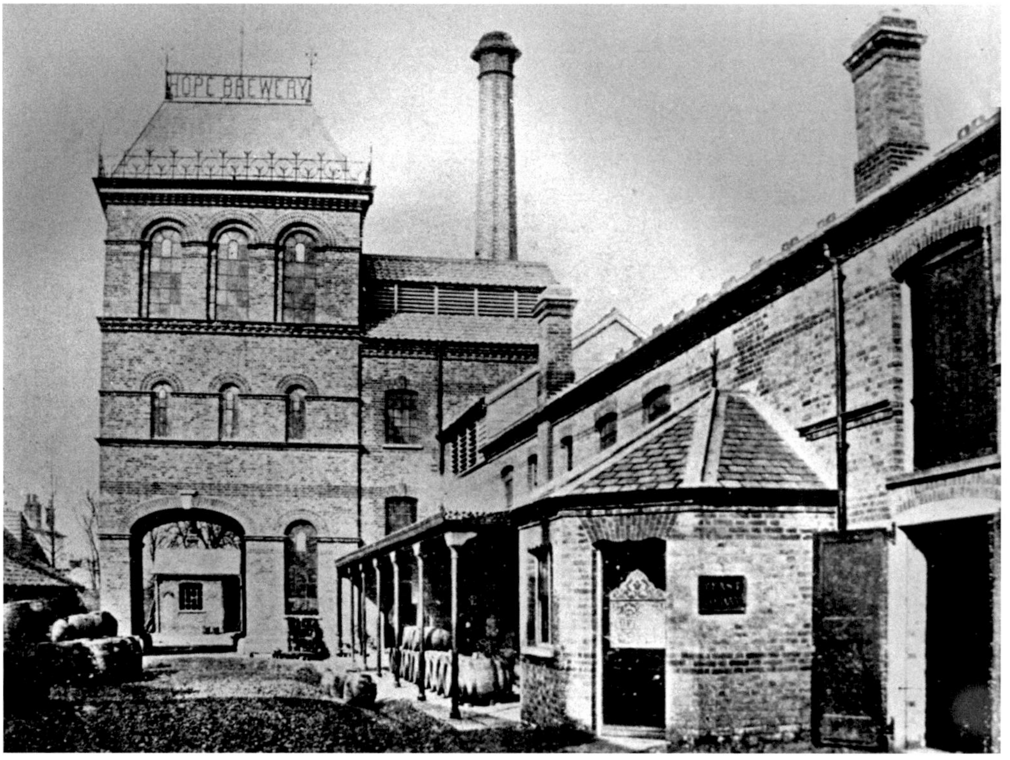

Baker's Hope Brewery, with a sign outside the office saying 'Yeast today'. Brewers sold their excess yeast to the public for a variety of uses, including breadmaking and as a skincare product.

was taken back by the Baker family, with William Baker – presumably the son of the brewery's builder – at the brewery for the 1851 census, and married to Cater's widow, Susanna.

By 1916 the brewery was being run by Herbert Baker, a lieutenant colonel in the British Army. His elder son, Lieutenant Herbert Norman Baker, died in August 1917 of dysentery, aged twenty-six, while serving with the 2nd Nigeria Regiment in German East Africa. The brewery and its twenty remaining tied houses were sold to the neighbouring firm of McMullen's on the last day of 1920.

Crown Brewery

The brewery Thomas Driver Medcalf opened in 1863 was down what is now Railway Place and the Saracen's Head, which opened in 1849, was its tap. In November 1864 Medcalf died, aged sixty-one, after being thrown from his chaise. The brewery passed to his brother, James Driver Medcalf, and James's son, Thomas Driver Medcalf junior.

Thomas junior was still advertising his 'genuine home-brewed ales' from the Crown Brewery in February 1884. By August 1885 the brewery had passed to Percy Hargreaves, who already ran a brewery in Abridge, Essex. Seven years later the Crown Brewery was described simply as the 'Abridge Brewery stores'.

Dolphin Yard Brewery

The Glove and Dolphin Inn, Maidenhead Street, was in existence in Hertford as the Glove in 1621. At the start of the 1840s Joseph Barnard was running a substantial brewery in Dolphin Yard.

In May 1842 the brewery at Dolphin Wharf was put up for auction. On 27 July 1844, at Hertford Court House, a hearing was held into the bankruptcy of Joseph Barnard 'of the Dolphin-wharf-brewery, Maidenhead Street, coal merchant, Brewer and Maltster'.

Duncombe Arms, Railway Street

Thomas South is recorded as brewing in Back Street in 1832. In 1845 his son William South bought No. 24 Back Street, and in 1849 Thomas South successfully applied for a retail licence for the premises, naming it the Duncombe Arms.

The 1861 census records William as a brewer and publican at the Duncombe Arms. He looks to have left the pub by 1864, and mentions of brewing end, though the brewery is shown behind the inn on a map of 1880.

Gillman and Gripper, Bircherley Green

John Gillman, a grocer and tallow chandler, and John Gripper, a maltster, were leasing a brewery on the west side of 'Butchers Green Lane' (Bircherley Street) in 1781, next to Lombard House, which they bought from Richard Alsager in 1785.

Moseley Gillman was still described as a brewer in October 1816. Soon after he sold the brewery to Robert and John Horner, who in turn sold it in July 1822 to John Christie and George Cathrow of the Hoddesdon Brewery.

McMullen's

Family tradition says that Peter McMullen began brewing in the rear of his father William's house in Back Street (now Railway Street), Hertford, in 1827, aged twenty-nine, illegally selling his beer out the front of the house. Certainly in 1848

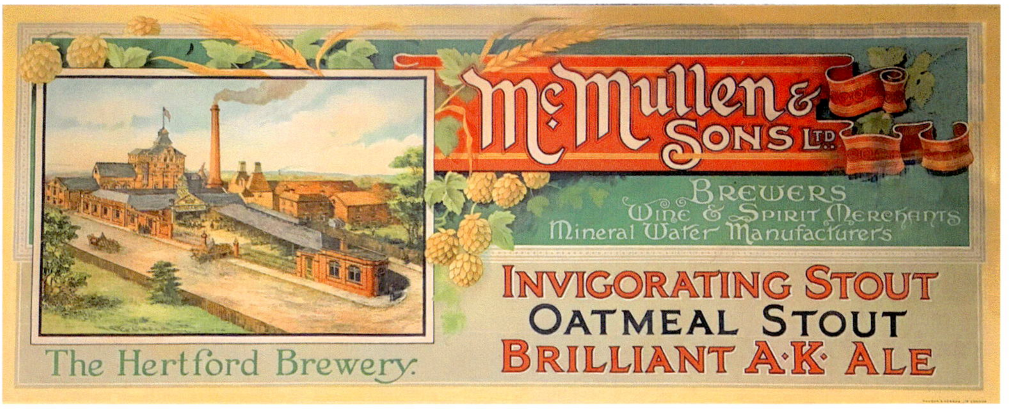

A showcard illustrated with the brewery in Hartham Lane. This opened in 1891 and was designed by William Bradford, who also designed the Fox Brewery in Bishop's Stortford and Wells' Lion Brewery in Watford. (Courtesy of Kevan Witt)

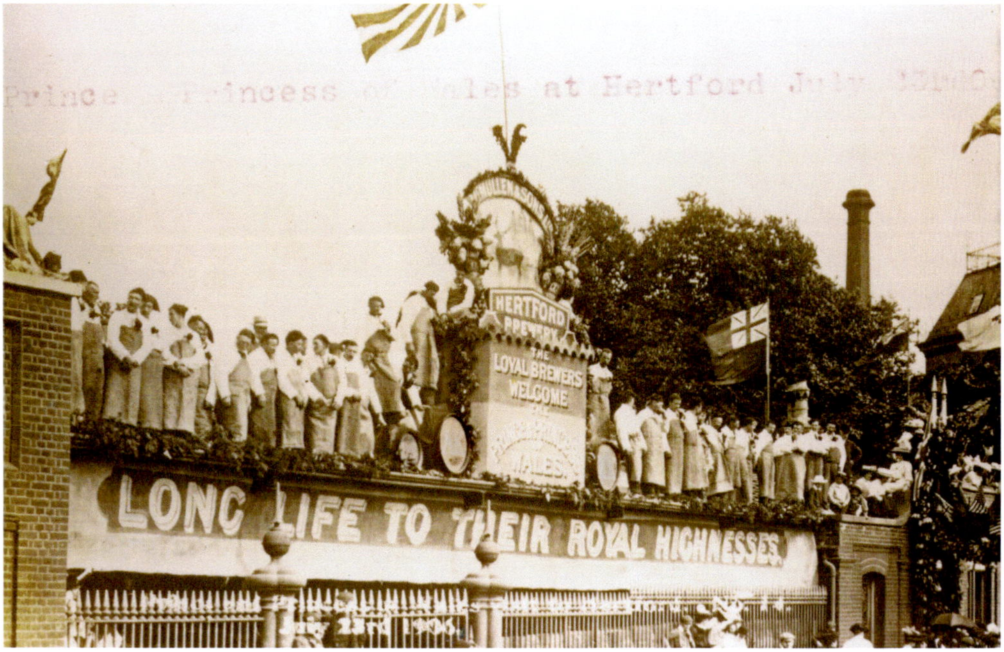

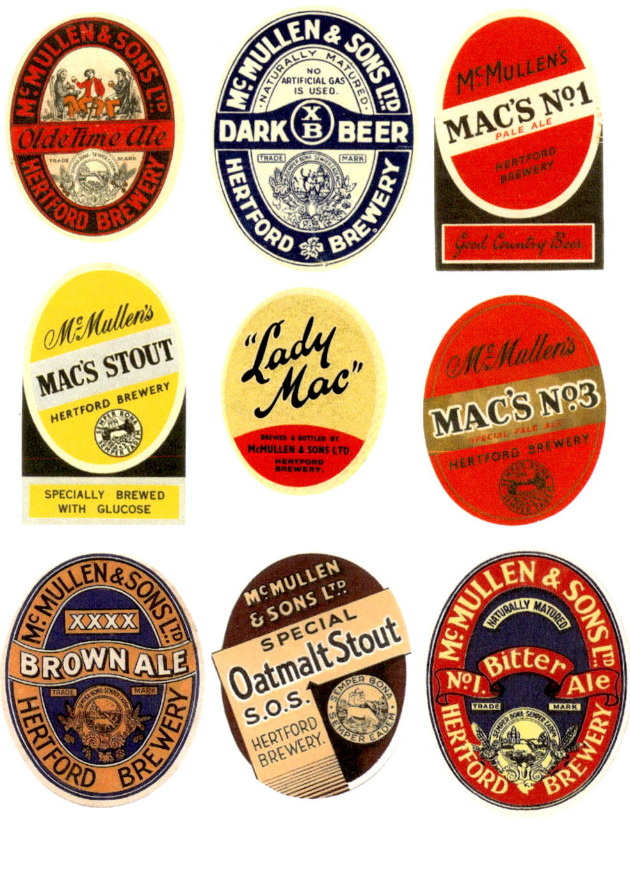

Above: Brewery workers, dressed in their traditional uniform of red stocking cap and apron, line up on top of the brewery wall in July 1906 to greet the Prince and Princess of Wales (later King George V and Queen Mary) on their visit to Hertford to open newly built schoolrooms at the Bluecoat girls' school. The presence of the Japanese flag is mysterious, unless it was to show solidarity with Japan in its war with Russia, which had ended the previous year.

Left: A selection of different McMullen bottle labels, including one for the curious 'Lady Mac' from the 1950s, aimed at female beer drinkers.

Peter McMullen placed an advertisement in the *Hertford Mercury* thanking the public for their patronage during the twenty years he had occupied the Mill Bridge Brewery, suggesting a foundation date of 1828.

Peter took advantage of the passing of the Beerhouse Act in October 1830, and paid two guineas (£2 2*s*) for a beerhouse licence at his home at Mill Bridge, naming the beerhouse after the new king, William IV. In 1832 he was listed as a brewer, beer retailer and cooper at Mill Bridge.

The McMullens were originally McMillans, from Scotland, who moved to Ulster. Peter's father, William, born around 1766, had moved from County Down to Hertford around 1790.

Peter McMullen bought his first tied house, the Greyhound, in Bengeo, in 1836. Around 1863 he retired, aged sixty-five. His two youngest sons, Alexander Peter, then twenty-eight, and Osmond Henry, then twenty-six, took over.

Late in 1863 the McMullens acquired a second brewing site, the Cannon Brewery in Princes Street, Ware. In April 1868, McMullens bought a third brewery, on the south side of Railway Street, and carried on brewing at the Bridge Brewery and in Railway Street and Ware for the next three decades.

A fourth brewery, Isaac Everitt's Star Brewery in Westmill Road, was acquired in November 1874.

Peter McMullen died aged eighty-two in April 1881. Alexander Peter's second son, Leonard, born in 1865, bought the Wimbledon Brewery in Wimbledon High Street, South London, in February 1888. In January 1889, the brewery burnt to the ground. Leonard then purchased the West Kent Brewery in Beckenham, Kent. The business ran until 1900, when the brewery closed.

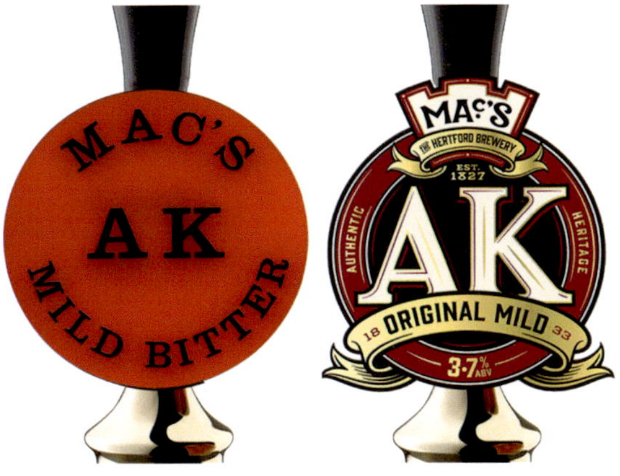

Two pumpclips for McMullen's AK, one from the 1950s, when the beer was described as a 'mild bitter', and one from 2019, when it was being called a mild, though originally in the nineteenth century AK was a popular name for a beer described as a 'light bitter ale' or 'family bitter ale' and sold for 1*s* a gallon, implying an abv of around 4.5 per cent. McMullen's claims today that it has been brewing AK since 1833, although the first advertisement for AK brewed by the company looks to appear only in 1903, when it was called a 'dinner ale'. The origin of the name remains unknown; brewers also made beers called AKK, XK, KK and other variants.

By 1889 the Hertford business was again outgrowing its capacity, and McMullens built an entirely new brewery on a triangular-shaped piece of land in Hartham Lane, designed by William Bradford, who was responsible for Harvey's Brewery in Lewis, Sussex, and the Hook Norton Brewery in Oxfordshire, among others. The new brewery opened in November 1891.

In 1897 the new firm of McMullen & Sons Ltd made its first takeover, the tiny Welwyn Brewery, with just one tied house. The following year McMullen's acquired George Henry Lee's brewery in Sun Street, Waltham Abbey, Essex, and eighteen public houses.

In 1907 the Epping Brewery, Essex, was taken over. On the last day of 1920, McMullens acquired their neighbours, William Baker and Sons' Hope Brewery at Old Cross.

In 2004, with beer sales falling, the Hartham Lane brewery was closed, the site sold to a supermarket in 2006, and a much smaller new brewery opened on the Old Cross site.

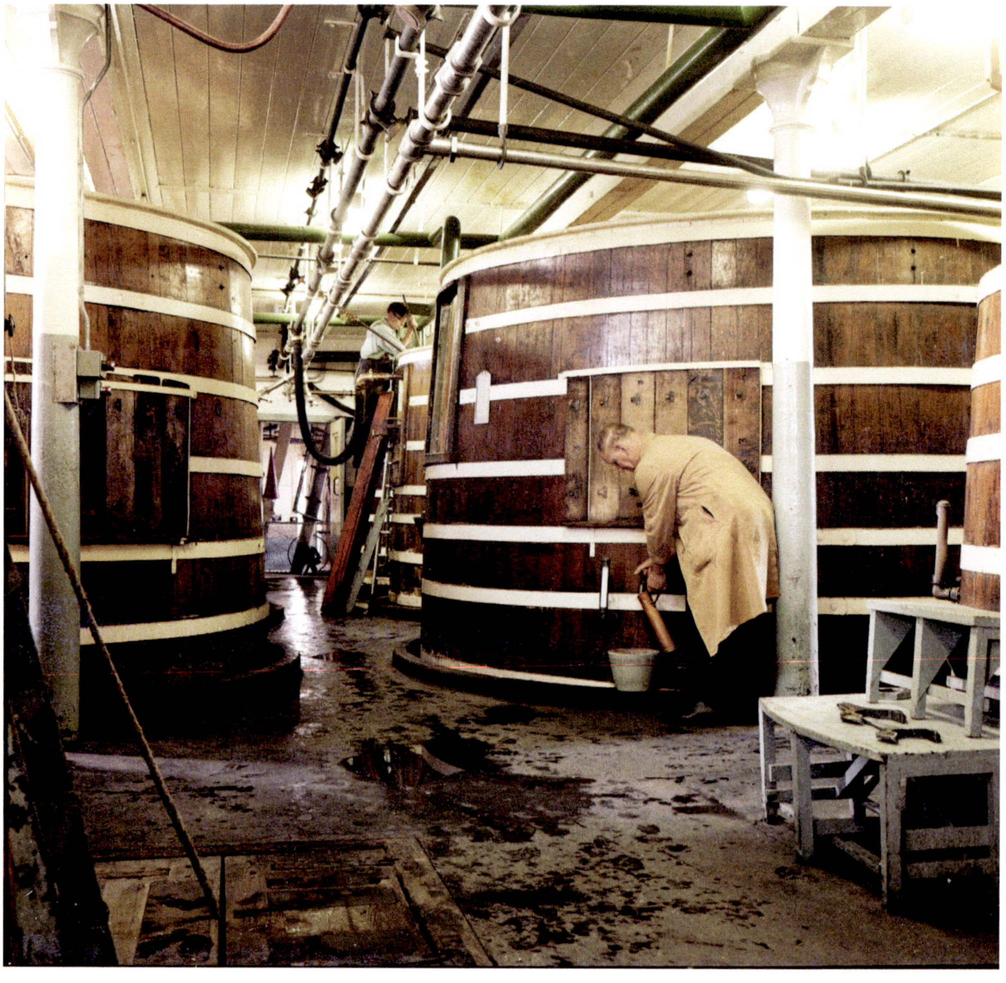

A brewer taking a sample from a wooden fermenting vessel at McMullen's Hartham Lane brewery.

Nicholl's West Street Brewery

Samuel Orgar Nicholls, born in Much Hadham in 1810, was running the Odd Fellows Arms, an evidently short-lived outlet in West Street, Hertford, in 1846. By the time of the 1851 census he had become a brewer living in West Street. Four years later he signed the lease on what was to become the brewer's house, a building in West Street built in 1719.

Samuel died in 1860, aged fifty, leaving the business to his eldest son, William Henry, aged twenty-five. William himself died in June 1879, at just forty-three, with his eldest son only fifteen years old. The business was carried on by his wife, Ellen. Under Ellen the business prospered, and in 1885 the brewery was rebuilt and re-equipped. A steam boiler was installed, which powered a 10-hp single-cylinder engine, both of which were still in operation when the brewery closed eighty years later.

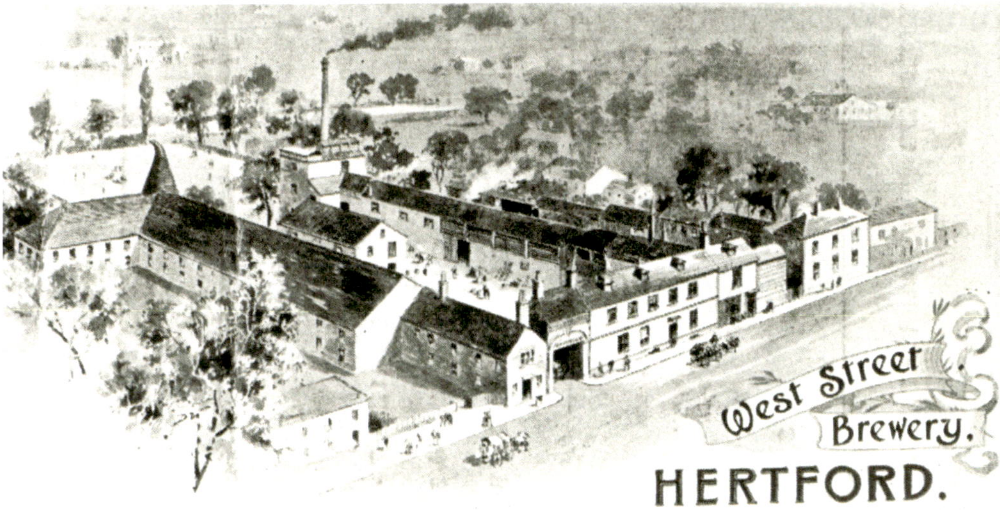

Above: Nicholls' West Street brewery, as illustrated in 1897. The brewery boiler house now lies beneath the Gascoyne Way dual carriageway. (Courtesy of Herts County Record Office)

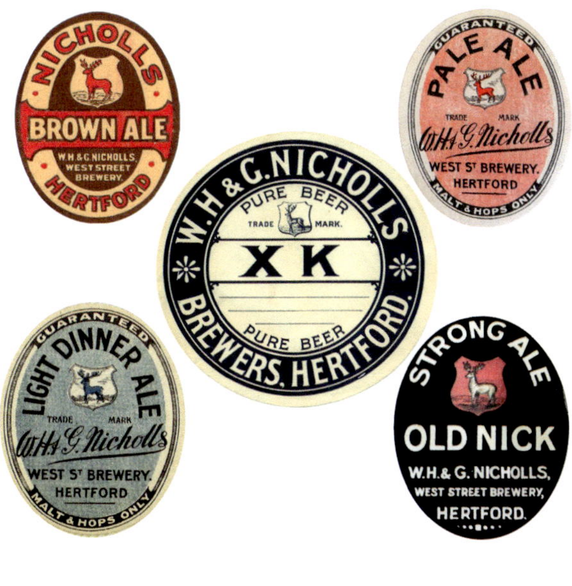

Right: Four Nicholls bottle labels and a cask label for XK draught beer. (Courtesy of Keith Osborne)

Ellen Nicholls retired from the business in March 1896, handing over control of the brewery and its four tied houses to her two sons, William Henry junior and George. William Henry Nicholls junior, a major in the Herts Imperial Yeomanry, died in September 1922 in London after a short illness. He was followed in the business by his son, William Henry III.

In the early 1960s brewing was down to perhaps once a month. The end came when a new ring road for Hertford was announced, with a route that went right through the brewery boiler house. Brewing stopped for good in November 1965, and the brewery was purchased by Hertfordshire County Council. On 31 March 1966, at the retirement of William Henry Nicholls III, the business was closed down. Shortly afterwards, the brewery was demolished, leaving only the brewer's house and the maltings standing.

Railway Street Brewery

William Manser, a Quaker, was listed as a brewer in 1822. His brewery was in Providence Court, between Fore Street and Back Street. Manser was still brewing in 1835.

However, the brewery appears to have been leased from 1838 by Thomas Driver Medcalf, the Hertford auctioneer and estate agent. Medcalf's lease on the brewery ended in December 1861, and in June 1863 he announced that he was now operating from the newly built Crown Brewery in Ware Road. His old brewery was taken by William John Palmer, a Hertford maltster, from 1 January 1862.

In December 1864, a fire almost totally destroyed the brewery. Palmer looks to have rebuilt the premises, but in April 1868 he announced that he had sold the business to McMullen & Sons.

St Andrew Street Brewery

John Ireland, formerly of the Ship, Old Cross, bought a malthouse and other property down what was then Newsells Lane, off St Andrew Street, in March 1774 for £500. By the time he died in May 1778, aged fifty-two, Ireland had built a brewery there.

His son Timothy continued the business, until in July 1785 he signed a lease on the brewery to Richard Flower and John Fordham. Flower, who was around twenty-three at the time, came from London, and had trained at Whitbread's Brewery in Chiswell Street.

Flower stayed at the brewery until May 1802, when he sold his interest in the business, with a tied estate of at least fourteen or fifteen pubs, for £6,000 to two maltsters from Baldock, Thomas Wells Fitzjohn and George Fitzjohn.

His son, Edward Fordham Flower, eventually settled in Stratford-upon-Avon, where in 1831 he founded a brewery of his own, which ultimately became part of the Whitbread empire.

In October 1817 Thomas Fitzjohn sold the lease of his brewery and its pubs to John Moses Carter, five times mayor of Hertford and political agent for the Marquess of Salisbury.

The brewery eventually grew to entirely block Newsells Lane. By 1831 it was described as a 'private way' to the brewery, and it became known simply as Brewery Lane.

In 1842 Carter sold the brewery to John Back, from Surrey, and his partner Robert Hunt, from Stamford, Lincolnshire, for £28,000. Just four months after buying Carter's brewery, however, Back and Hunt agreed with the owners of the much larger

Hoddesdon Brewery to buy that, and its 104 tied houses, as well. Eventually, in August 1844, Back and Hunt and their new partner, Peter Christie, closed Carter's brewery.

Wickham, Mill Bridge

When Wickham's brewery on Mill Bridge was rebuilt, the tiles on the outside proclaimed it to have been founded in 1828. John W. Ireland is listed as a brewer at Mill Bridge in a directory of 1829. That same year the brewery was acquired by Edward Wickham, born in Thornham, north Norfolk, in 1793.

Edward's oldest son, William, born in 1830, took over his uncle John Long's brewery in Ware High Street in July 1865, and for the next thirty-four years the two Wickham breweries had an agreement that they would not sell to customers in each other's home towns.

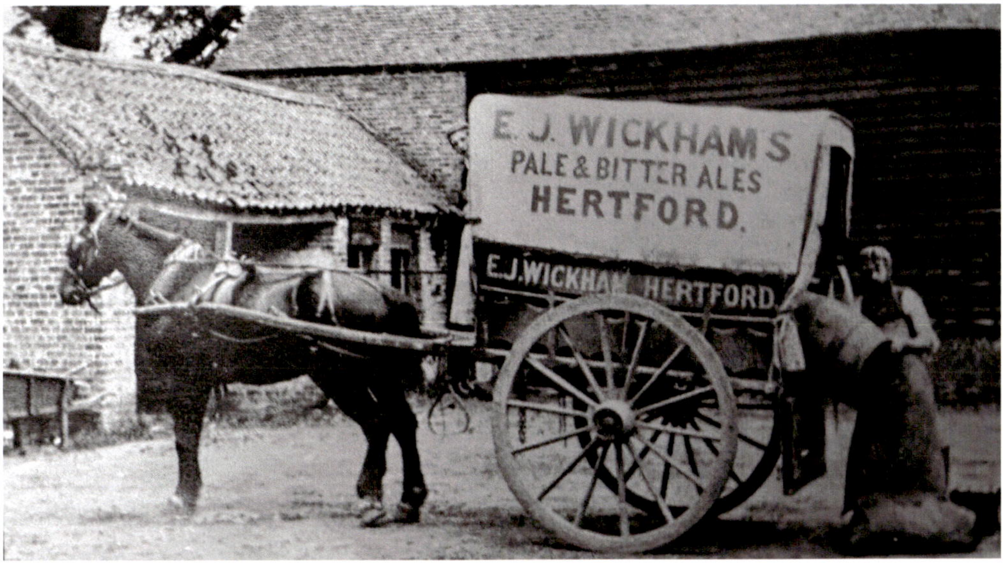

Above: A dray being loaded in the yard at E. J. Wickham's Mill Bridge Brewery.

Right: A label and stopper strap for Wickham's Mill Bridge Brewery. (Courtesy of Keith Osborne)

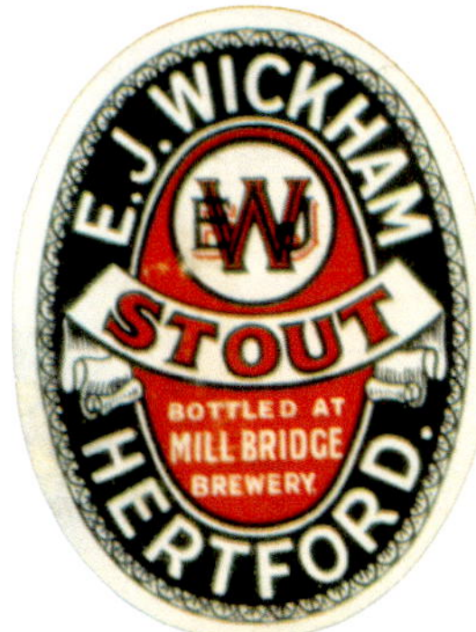

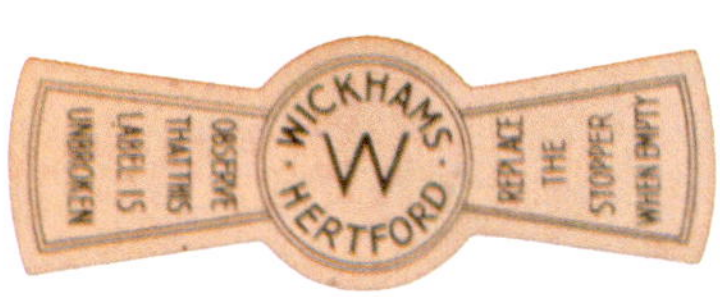

In 1874 Edward, then eighty-one, passed the Mill Bridge Brewery to his other son, Edward John, born in 1844. By 1938 the brewery was being run by Edward John's sons, Edward, then sixty-seven, and Harold, fifty-eight. That year they accepted an offer of £20,000 for the Mill Bridge Brewery and its tied houses from the Biggleswade brewers Wells & Winch.

Thomas Wing, Castle Street Brewery

Thomas Wing senior had a cooper's shop in Castle Street, where, by 1858, he had apparently started brewing. The business passed to his son, Thomas junior, and from 1864 advertisements appeared regularly for 'T. Wing's genuine home-brewed ales', at first from the Castle Street Brewery, then, more grandly, from the Castle Brewery.

In April 1881 Wing was declared bankrupt. He died in December 1884, aged fifty-one.

Young's Brewery, Fore Street

Joseph Catlyn, or Catlin, was leasing a maltings and a property 'lately converted to a brewhouse' on the corner of Sow Lane (now South Street), off Fore Street, in 1702. He died in 1729, and the next known brewer on the site was John Walter, in 1736. Thirty years later, Walter was leasing the brewery to Richard Boor.

In 1770 James Boor was leasing the brewery to John Pycroft of Greenwich, brewer. Boor died in 1776, and the following year his sister, Elizabeth, was leasing the brewery

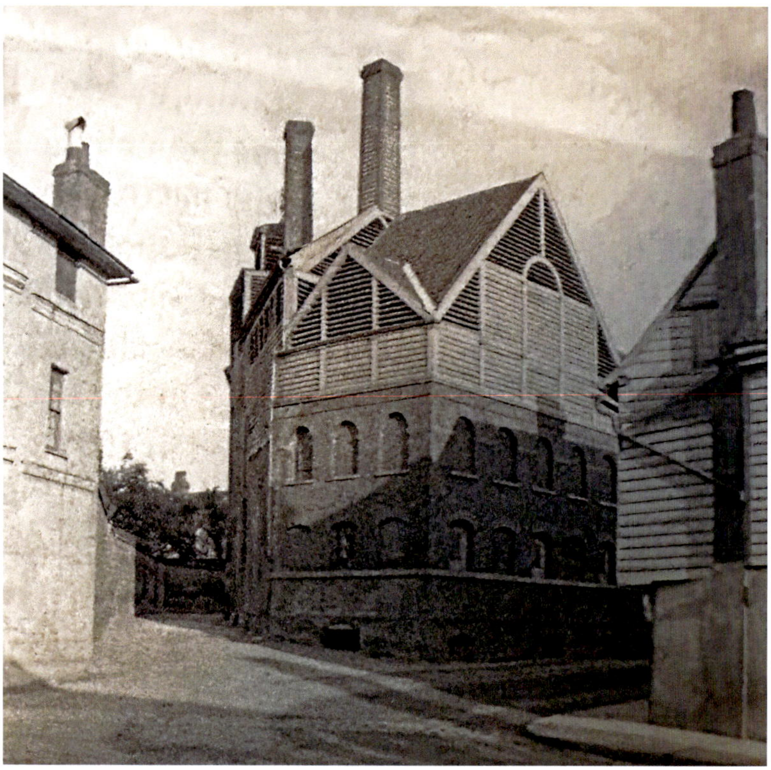

Young's Brewery in Fore Street, Hertford, had a history of brewing dating back 195 years when it closed in 1897.

to Robert Silversides and Robert Young. Noah Young, Robert Young's son, was running the brewery by 1791.

On Noah's death, aged forty-eight in 1829, the brewery passed to Noah's son, Noah Robert. When Noah Robert died in August 1842, aged only thirty-four, his brothers, Benjamin and Charles Young, who were aged twenty-six and twenty respectively, took over.

By 1870 the brewery had thirty-six tied houses. In October 1893 Benjamin Young was seventy-seven, and it was announced that the brewery and its tied houses had been sold to Pryor Reid of Hatfield. Pryor Reid carried on brewing at Hertford until 1897.

HITCHIN

Bedford Arms Brewery

After Arthur Searle Hoye disposed of his brewery in Tilehouse Street (see below), he moved to Bedford Road, where he was living in 1861. In around 1865 he opened a beerhouse in the street, calling it the Bedford Arms, and began brewing there.

Hoye carried on brewing at the Bedford Arms until his death in December 1881, aged sixty-four, when brewing at the beerhouse looks to have stopped.

Hitchin Brewery, 5 Sun Street

John Crabb was already a brewer, presumably at the premises later known as 5 Sun Street, next door to the Sun Inn, when he married his second wife, Hannah Ewesden, in 1753. The business, which by now had some twenty-six tied houses, was being run in 1826 by his great-grandson, Henry Crabb, born in 1796, and John Marshall.

Henry Crabb died of consumption in 1830. In 1832 Marshall took a new partner, Joseph Margetts Pierson junior, whose father had started a bank in Hitchin with the Crabb family. By the early 1840s the brewery was producing 5,500 barrels a year, while it supplied fifty tied houses. However, Marshall became insolvent, and the Hitchin Brewery and all its tied houses were put up for sale in March 1841 at the Sun Inn, a dozen or so different local brewers dividing up the estate between them.

Joseph Margetts Pierson junior died in 1842, but his son, Joseph Margetts Pierson III, born in 1811, set up his own brewery business barely a hundred or so yards away between Bucklersbury and Tilehouse Street, under the name of the Bucklersbury Brewery. This ran until summer 1852, when Pierson decided to emigrate to New Zealand, and the brewery and its 'several' tied houses were put up for auction.

Lucas

Lucas family tradition says that they acquired the brewery on the corner of Sun Street and Tilehouse Street in 1709, when William Lucas II, second in a line of eight William Lucases, a Quaker farmer and maltster, went into partnership as a brewer with his brother-in-law, Isaac Gray.

The reality appears slightly more complicated. Lucas and Gray may have been brewing originally at a brewhouse on the north side of Tilehouse Street. The pair were running a brewery at the Red Lion in Bucklersbury some time after 1727 – possibly after 1734. By 1748 they had acquired the brewery in Sun Street formerly owned by Edward Draper, who had been brewing at the Angel and Chequer in Sun Street from at least 1698, after the Draper family were unable to pay off a loan to William Lucas of £1,100.

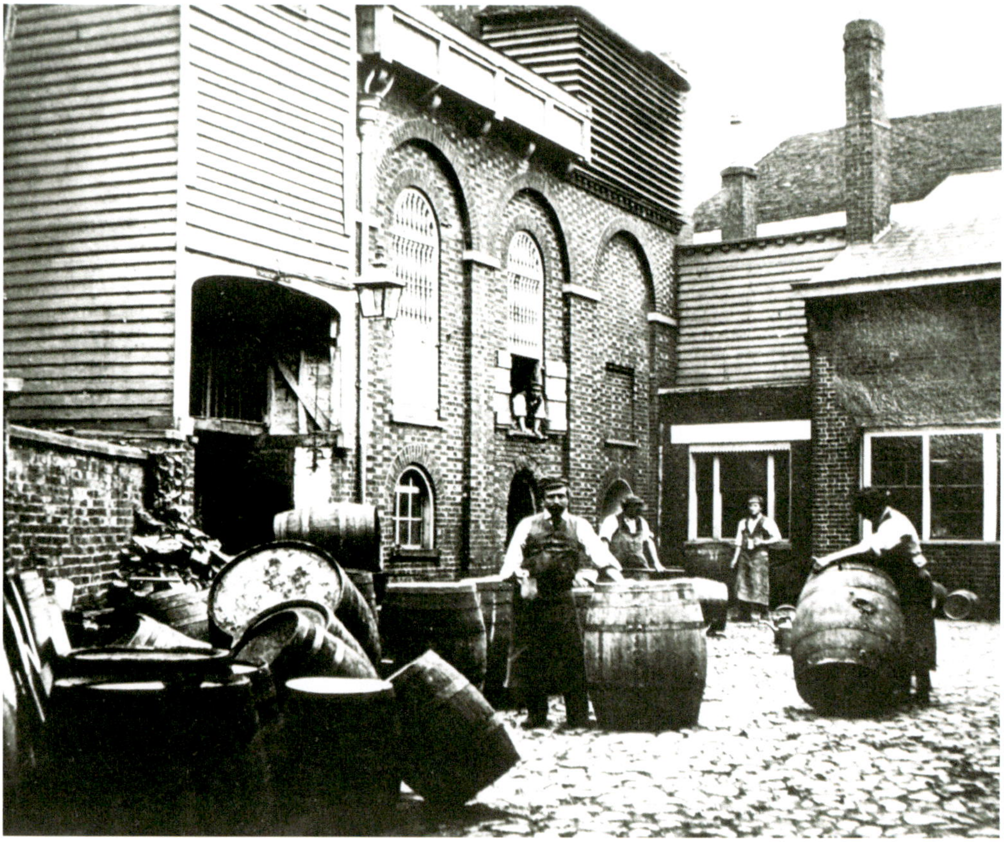

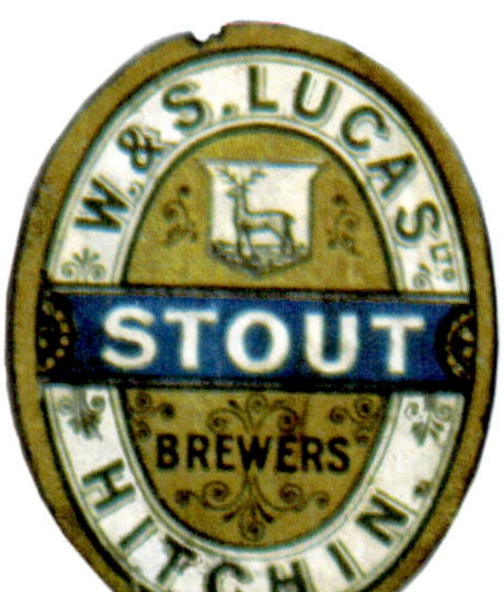

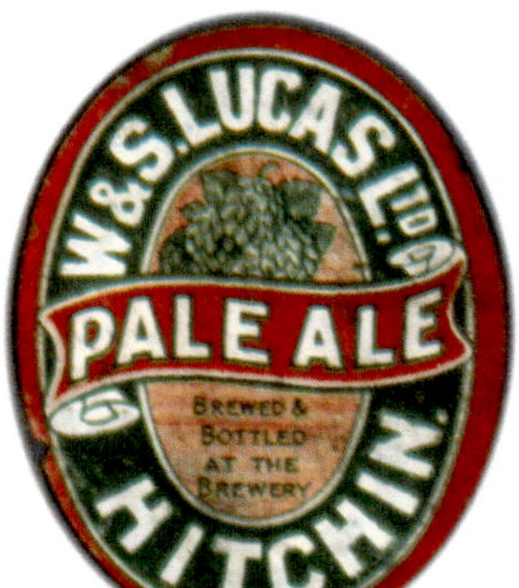

Above: The yard at Lucas's Brewery in Sun Street, Hitchin.

Left: Two bottle labels from the Lucas's Brewery, one with the 'hart in a ford' logo the company shared with several other Hertfordshire brewers. (Courtesy of Keith Osborne)

The business continued under a line of William Lucases, slowly growing, so that around the 1850s the Sun Street brewery had at least forty tied houses. The family's Quaker scruples always brought tensions to their relationship with the way they made money. William VI's eldest son, William Lucas VII, joined the brewery in 1853, when he was twenty-one. His worried father wrote in his diary: 'Oh Lord, grant … that our eldest son, who is now coming home, may be preserved from the peculiar snares of our business, that he may not have his moral sense dulled by the impurity with which it abounds.'

William VII died in 1914, aged eighty-two, and his cousin, Samuel, in 1919. The following year, in September 1920, the company and its fifty-two surviving public houses were acquired by the J. W. Green Brewery in Luton. The brewery continued in operation until June 1923, when it was announced that brewing was to cease.

Peacock Brewery, Queen Street

William Young was landlord and brewer at the Peacock in Back Street (later Queen Street) in 1845, and still brewing at the Peacock in 1853. By 1859 Henry Seaman was running the Peacock. Then in October 1864 the licence of the Peacock was transferred from Seaman to William Chapman.

Chapman appears as a brewer at the Peacock in *Kelly's Directory* of 1906. That October, however, the Peacock was taken over by the Herts and Essex Public House Trust (later Trust Houses), and brewing stopped.

Port Mill Lane Brewery

John Bradley Geard, born in Hitchin in 1783, was brewing in Port Mill Lane by 1838. The brewery, 'held under lease by Mrs J. B. Geard', was put up for auction in July 1846. In December that year its tied estate, six beerhouses from Stevenage to Ashwell and three pubs, was auctioned off.

It seems to have been acquired by J. and W. Lewin, but looks to be occupied by George Deards, listed as a brewer in Port Mill Lane in 1851 and 1854. C. King was brewing in Portmill Lane in 1861.

The brewery was occupied from 1862 by William Lewin. In November 1864 Lewin advertised that he had moved to the Triangle Brewery, Hitchin. The Triangle was at the foot of Bridge Street. Lewin did not last there very long: in May 1867 Lewin's Triangle Brewery and effects were sold off by Jackson's, the Hitchin auctioneers, for £61 18*s*. All the same, Lewin was listed as a brewer in Bridge Street, Hitchin, in 1874.

Sun Brewery

The Sun, which gave its name to Sun Street, is frequently stated to date back to at least 1575. There is, however, no definite evidence of the Sun Inn until 1650. An inventory in 1754 valued the inn's brewhouse at £80 2*s* 6*d*, around 30 per cent of the total value of the inn.

William Hill, who ran the Sun for almost fifty years, dying in 1888, is listed as a brewer in 1874. He was followed at the inn by its former manager, Edwin Logsdon, then thirty-eight. Logsdon's son, Ted, trained as a brewer in London and returned to run the inn's brewery and mineral works. Logsdon retired from the inn in July 1900, to

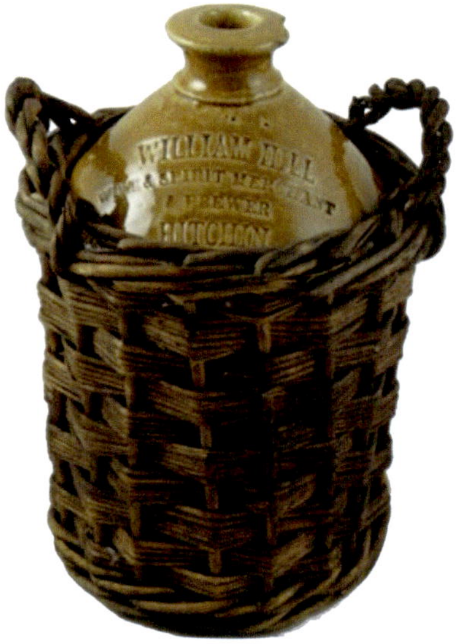

A flagon from the Sun Hotel and Brewery bearing the name of William Hill, the inn's owner until his death in 1888. Unusually, the flagon still has its wicker basket intact. One of the Lucas family made the excellent pun that it was 'the Sun that tips the Hills with gold'.

be followed by George Rochford. Rochford died in January 1908 after slipping on the frozen grass of the hotel lawn, and the brewing side of the business seems to disappear after his death.

Tilehouse Street Brewery

Arthur Hoye was a jobbing brewer living in Back Street in the 1851 census, listed as next door to William Young at the Peacock Brewery and probably working for Young. Later in the 1850s he was brewing in his own right in Tilehouse Street. His business was then acquired by William Hall junior, who described himself in 1862 as 'Tilehouse Street Brewery, Late A. Hoye'.

In March 1875 advertisements appeared for the sale of the Tilehouse Street Brewery's equipment, on the instructions of William Hall, 'who is relinquishing the business'.

HODDESDON

Christie

The Hoddesdon Brewery sprang out of the brewhouse at the Thatched House in the High Street, immortalised by Izaak Walton in his book *The Compleat Angler*, first published in 1653, when he declared the inn 'very remarkable' for its ale.

In around 1700 the Thatched House was acquired by William Plomer, from Amwell. Under William, who died in 1728, and his son Robert the brewhouse at the inn developed into a thriving brewery business, and the inn itself disappeared.

After Plomer's death, the business was acquired by Rene Briand, initially in partnership with Nathaniel Dallowe. Briand died in 1781, and the brewery passed to William Whittingstall, member of a family of rich Hertfordshire millers. William died in 1803 and the brewery was sold to William Christie, born in Wemyss, Fifeshire, in 1744, and his business partner, George Cathrow.

Christie died in 1811, and his eldest son, John, born in or before 1787, carried on in partnership at the brewery with Cathrow. John Christie died in July 1831, and George Cathrow passed away not quite eleven years later, in February 1842, aged sixty-seven.

In February 1843 Christie and Cathrow's executors agreed to sell the brewery and its 104 tied houses to John Back from Surrey and Robert Hunt from Stamford, Lincolnshire, for £86.

Back and Hunt had already, a few months earlier, taken over the Newsells Lane brewery and its forty-two pubs and beerhouses in Hertford from John Moses Carter. A short while after the takeover of the Hoddesdon Brewery they agreed to take in as a third partner Peter Christie, nephew of William Christie, who had been working at the brewery for some years.

Peter Christie retired in 1858, replaced at the firm by his son, Charles Peter Christie, who ran the brewery until his death in 1898, aged sixty-nine. Charles Peter Christie's grandson, John Fairfax Christie, was a captain in the 1st Battalion, Hertfordshire Regiment and wounded twice and awarded the Military Cross for his actions during the last great German push in March/April 1918. (His commanding officer was Lieutenant Colonel Eric Phillips of the Royston Brewery, who was taken prisoner by the Germans during the same offensive.)

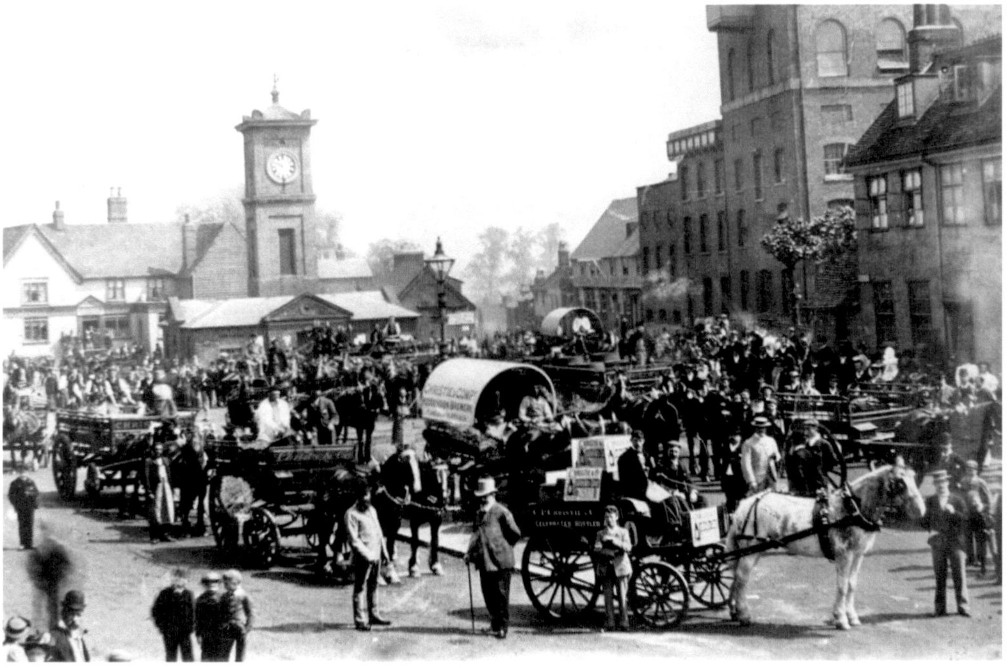

Christie drays drawn up in front of the brewery before taking part in a parade to Hertford, possibly the Bank Holiday parade of June 1889. The man with the cane and top hat front and centre is very likely C. P. Christie.

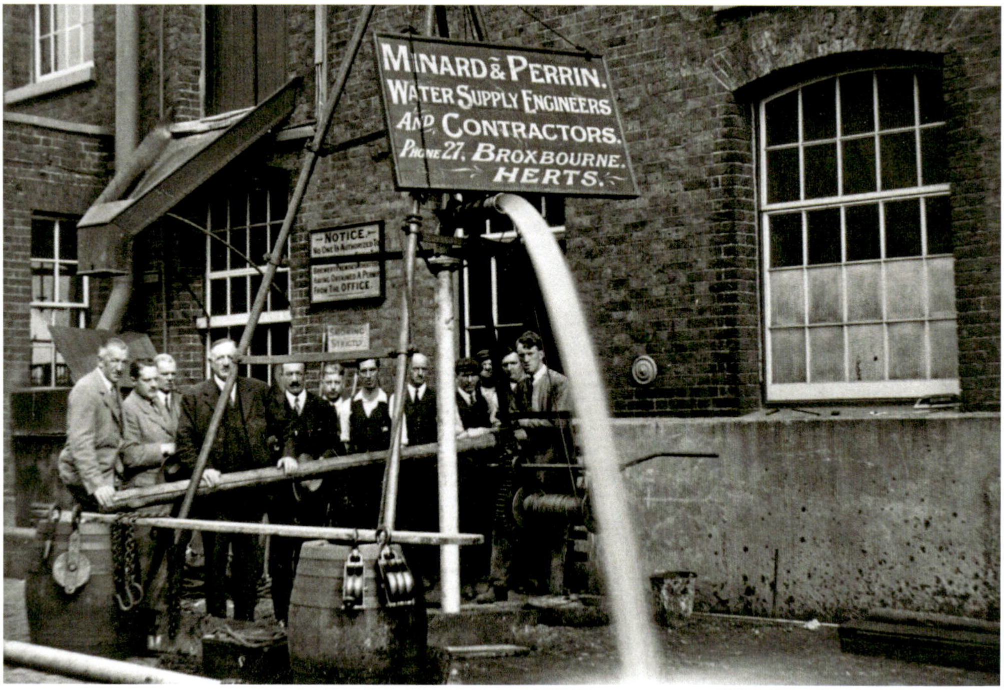

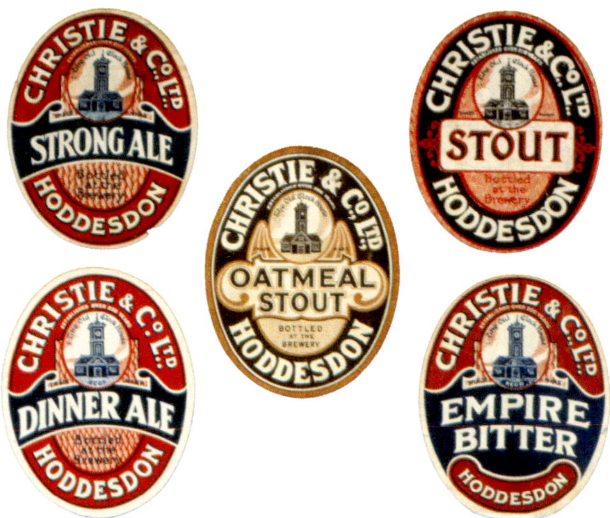

Above: Boring for water at the Hoddesdon Brewery, *c*. 1925. Captain John Christie MC is second from left. (Courtesy of Hoddesdon Library)

Left: Five Christie's labels bearing the company's Hoddesdon clock house trademark. (Courtesy of Keith Osborne)

After the war Captain Christie was sent off to learn the trade at a brewery in Kent. In April 1921 he came back to the Hoddesdon Brewery, rising to become under-brewer and a director of the firm. But the captain suffered terrible headaches from the wound he had received in the trenches. In October 1927 he shot himself in the head, dying in hospital some hours later.

A month later it was announced that the brewery and its 153 tied houses had been sold to the Cannon Brewery Co. Ltd of St John Street, Clerkenwell, for £501,800.

ICKLEFORD

John Foster

John Foster was born in 1813 in Holwell. By 1861, Foster had become a beer retailer in Ickleford. Probably from the start he was a brewer as well, at his home, Mayfield Villa, in Arlesey Road.

In 1891, John Edward Foster, brewer, aged twenty-five, John's grandson, was living at The Brewery. The brewery was still listed in 1898. By 1901, however, John Edward Foster had moved to Stevenage, and was working as a plasterer.

KIMPTON

Lion Brewery

Joseph Kingsley was born in Pirton in 1802. In 1828 he married Eliza Bruton, daughter of a Kimpton landowner, Christian Bruton. In 1832 Joseph came into property in Kimpton High Street formerly owned by Christian and turned it into a brewery.

He was certainly brewing by August 1836. That same year he bought the Melbourne Arms (later the Nelson) at Marford, near Wheathampstead.

The business hit financial problems, Kingsley apparently fled to Jersey with his family to avoid his creditors, and in 1848 the brewery was leased by the mortgagees (some being third mortgagees) to Kingsley's former manager, William Coleman, and Coleman's brother-in-law, John Sibley. In 1855 Sidley looks to have left, and Coleman was joined by Joseph Kingsley's younger son, Henry.

The firm continued as Kingsley and Coleman until Coleman died in 1873. In 1892 Henry was pulled under by his debts, and the brewery and nine tied houses were put

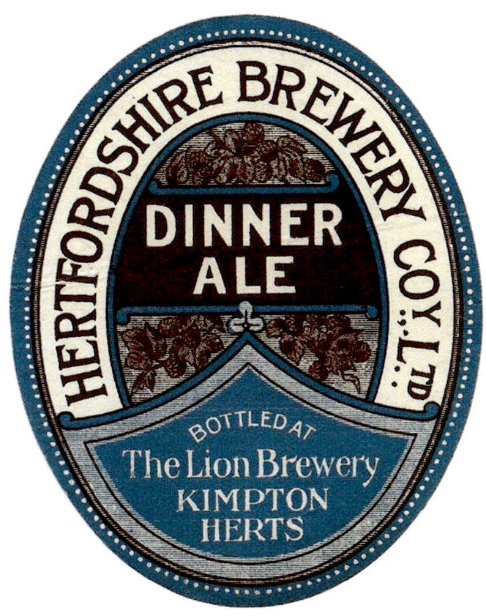

A label for dinner ale from the short-lived Hertfordshire Brewery Company of the Lion Brewery, Kimpton, active *c.* 1900. (Courtesy of Keith Osborne)

up for sale. Pryor Reid of Hatfield bought two pubs and two beerhouses, but the rest failed to sell. Eventually, in 1893, the brewery was acquired by William George Hornsey, a speculative builder from West Ham.

The business continued as Hornsey & Co., the Lion Brewery under the charge of an experienced brewer-manager, Henry Boon Clark, who had learned to brew at his family's farm, Hyde Hall, Sandon. But in November 1900 Hornsey sold the brewery to a new concern, the Hertfordshire Brewery Company Ltd.

Cyril Herbert Brass, twenty-nine, former brewer and manager of Clinch's Brewery in Whitney, Oxfordshire, was appointed managing director at the Lion Brewery. However, the Hertfordshire Brewery Company collapsed in November 1901, and a receiver was brought in.

Cyril Brass stayed on in Kimpton, and in June 1902 a new company, Brass and Abbott Ltd, took over the Lion Brewery. The brewery was put up for sale at the Mart in London, again in June 1904. The purchaser was a Kimpton man, Charles George Chalkley, aged thirty-nine, whose father was a wealthy builder in the village. The brewery was still running in 1908 but seems to have closed in 1909.

KINGS LANGLEY

Kings Langley Brewery

The Kings Langley Brewery seems to have grown out of a wine tavern called the Crown, which was being kept by James Goodwin in 1633. The tavern had apparently changed its name to the Rose and Crown by 1668. It passed down through three more generations of Goodwins, and at some time a public brewery began at the Rose and Crown. Late in the nineteenth century, the brewery was claiming a foundation date of 1720.

James Goodwin's great-granddaughter, Sarah, married a Kings Langley attorney named Thomas Groome in 1761. In 1767 Sarah gave birth to a son, named Thomas Goodwin Groome. Just four years later her husband died. Sarah herself died in February 1783, aged fifty-one, when her son was only sixteen. She left all her property to him, including fifteen inns and pubs.

In October 1801 Thomas married Ann Peacock Smith of Abbot's Langley, and they had a son, John Andrew Goodwin Groome, born in September 1805. Thomas died less than four years later, in March 1809, aged just forty-one. Control of the business passed to his widow until her death, aged fifty-one, in 1828, and then to her and Thomas's son John.

In the early 1850s John Andrew Groome was in partnership at the brewery with his brother-in-law, Richard Huntley. Their partnership was dissolved in November 1853. Groome had married Margaret Huntley in 1836. When Groome died in January 1864 the brewery was carried on by two of his sons, the eldest, John Edward Groome, born in 1837, and the third son, Henry Abbott Groome, born 1843, as Groome and Brothers. Henry died at Hastings in August 1869, and John Edward Groome carried on the business on his own.

By the 1890s the brewery, with its ten-quarter plant, was selling 4,500 barrels a year, made with water from its own well on the premises. In August 1897 Groome, who

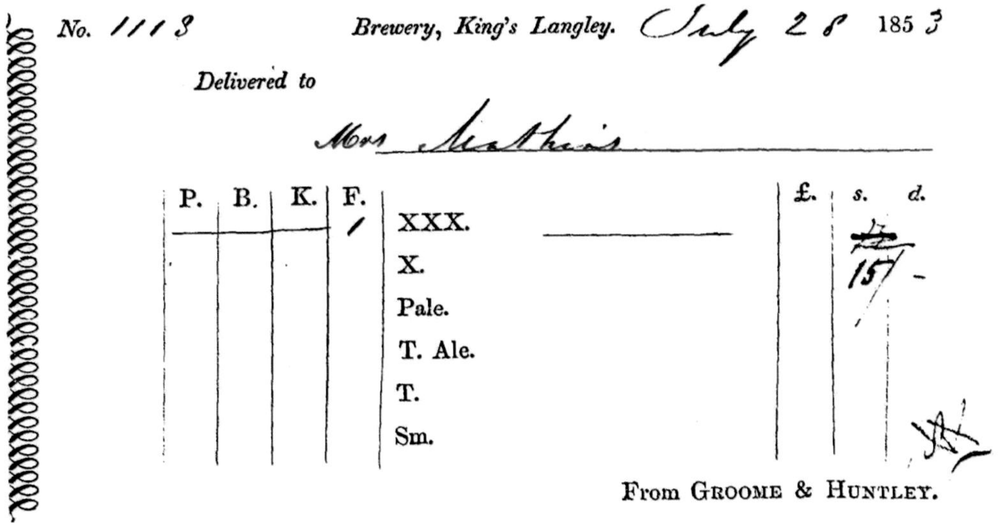

No. 1118 Brewery, King's Langley. July 28 1853

Delivered to

Mrs [illegible]

P.	B.	K.	F.		£.	s.	d.
			1	XXX.		15/-	
				X.			
				Pale.			
				T. Ale.			
				T.			
				Sm.			

From Groome & Huntley.

A receipt from the Kings Langley Brewery from 1853, when it was being run by John Andrew Groome and his brother-in-law, Richard Huntley. (Courtesy of the Brewery History Society)

was now sixty, living in Windsor and with no sons to succeed him, announced he was putting the entire business, including thirty-two tied pubs and beerhouses, up for sale at the Auction Mart in London. Strenuous bidding pushed the price up from £20,000 to £64,000, when the brewery and its tied estate fell to Benskin's of Watford.

The brewery was eventually demolished, though the yellow-brick brewer's house, built in 1830, still stands. John Edward Groome died in May 1890, and was buried in Kings Langley churchyard.

KINGS WALDEN

Henry Pinks Arnold

Henry Pinks Arnold was described as a farmer only at Medlow House, Coleman's Green, in the census of 1851, but that year he built a beerhouse, the Rose and Crown, in Wheathampstead. In 1855 he was a 'farmer and brewer'.

Arnold died around 1870, aged fifty-four or so, and the business was carried on by his widow, Mary, until her death in 1890. Henry and Mary's daughter, Mary Ann, and her husband, then Joseph Farr, ran the brewery until about 1894, but it looks to have ceased around 1896.

Kings Walden Brewery

The Fox inn at Frogmore was being run by John Buck in 1841, when the census describes him as a brewer at Frogmore. In 1855 Buck bought a beerhouse at Darley Hall, the year after the death of its owner, Samuel McDonnell. Eventually the Darley

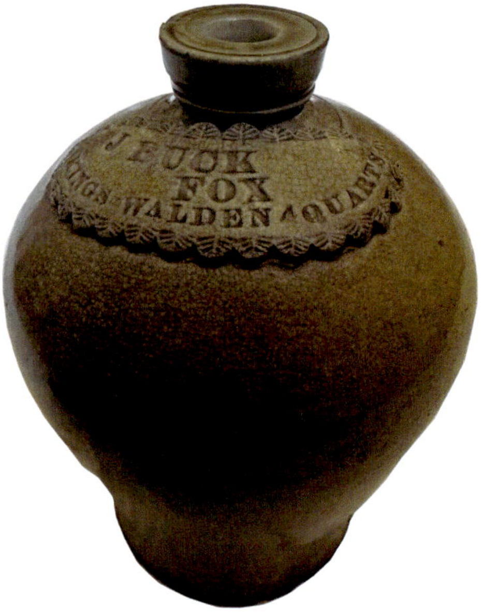

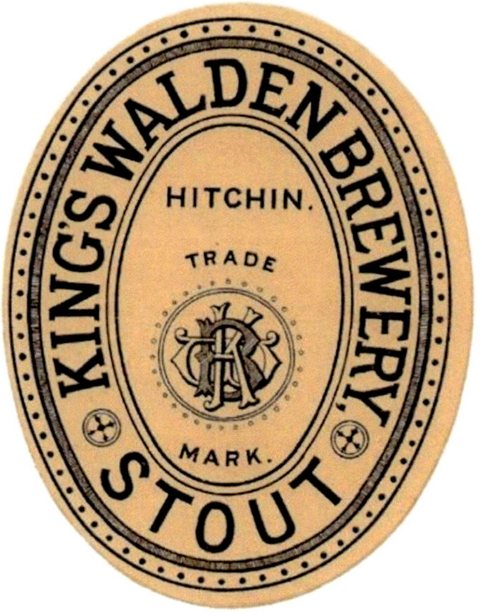

Above left: A slab-seal gallon jar from no later than 1871, used by John Buck, brewer at the Fox in Kings Walden.

Above right: A bottle label issued by the short-lived Kings Walden Brewery, *c.* 1890. (Courtesy of Keith Osborne)

Hall beerhouse, too, became known as the Fox. In 1868, John Buck married Samuel McDonnell's wife, Mary McDonnell. Buck died in 1871, and Mary, now fifty-eight, took over running the brewery, the farm and the Fox at Frogmore.

Mary handed over the licence at the Fox to William Walsingham in December 1883. Walsingham was still at the Fox in 1886, but soon after the premises were taken over by a partnership consisting of Frederick William Fellowes, born in Beighton, Norfolk, in 1856, his brother, the Reverend Henry Cecil Fellowes, personal chaplain to William Hinds, owner of Kings Walden Park since 1885, whose estate included the Fox at Frogmore, and Hinds himself. The trio then turned the premises into the Kings Walden brewery.

The Fellowes brothers' sister, Pleasance, had married Edward Jesser Coope, only son of Octavius Edward Coope, a partner in Ind Coope, and Frederick used his connection to Edward to train as a brewer in Burton.

William Hinds died in April 1890, leaving all his property to his wife, Rebecca Briggs Hinds. Nine months later, in January 1891, it was announced in the *Brewers' Journal* that Frederick Fellowes, the Reverend H. C. Fellowes and R. B. Hinds, 'under the style of the Kings Walden brewery, Kings Walden are dissolving their partnership'. In August the entire brewery plant was put up for sale.

Henry Timpson

Henry R. Timpson, born in Maidstone, Kent, around 1808, was listed as a retail brewer in Whitwell – probably, in reality, in Breachwood Green – in 1839.

Henry Timpson was a brewer and landlord at the Bricklayer's Arms in Breachwood Green in 1850. He continued to appear in trade directories as a brewer until at least 1870. By 1874, however, although still farming, he had evidently stopped brewing for sale.

LITTLE HADHAM

Nag's Head

The Nag's Head at Hadham Ford was opened in around 1861 by George Feltham, brewer, of Little Hadham and he promptly leased it and its brewery to George Stracey, aged thirty-five.

Stracey decided to retire in 1893 and the four-and-a-half-quarter brewery and the Nag's Head alongside were bought by Ernest Alfred Dalby, aged twenty-three.

By 1899 Dalby was in partnership with Francis Michael Tordiffe, aged twenty-one, formerly a brewer at the Fenstanton Brewery in Huntingdonshire. However, in summer 1899 Dalby left to run the Eaton Socon Brewery in Bedfordshire.

Tordiffe entered into a new partnership with Bernard Springett Springett, aged twenty-five, also from Kent.

By 1902 the brewery was being run by W. H. Drake. In around 1905, Drake was joined in partnership by Andrew Murray Lawson, the former manager at Rayment's Brewery.

Lawson was in sole control at the brewery in 1908. In 1912 he sold out to his old employer, Rayment's, and the brewery closed.

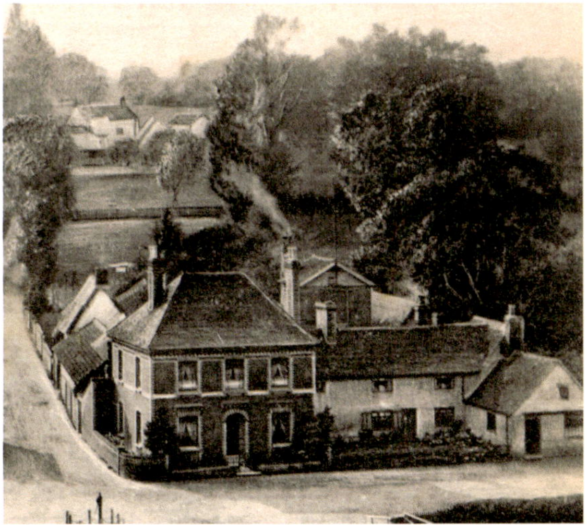

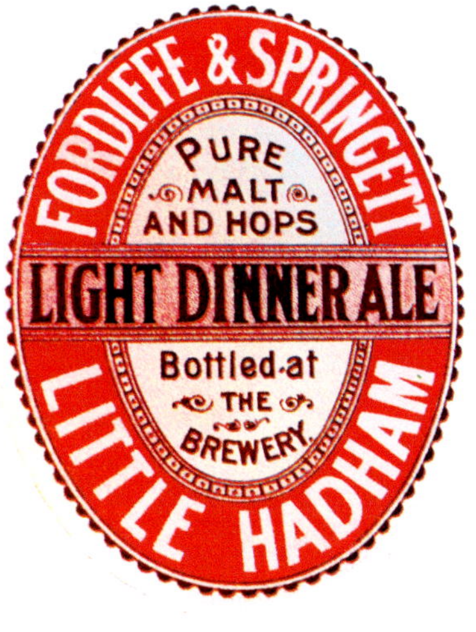

Above left: The brewery at Hadham Ford, visible behind the Nag's Head pub.

Above right: A mystery label purporting to be from 'Fordiffe and Springett' at the Little Hadham Brewery. There is no doubt that the partners in the business were Francis Tordiffe, with a 'T', and Bernard Springett. Was the label a printer's error or part of a scam? (Courtesy of Keith Osborne)

LONDON COLNEY

Colne Brewery

In March 1868 Thomas Lavender was brewing at the Colne Brewery, London Colney. In 1871 Frederick Lavender, twenty-four, was a brewer and maltster living in the village.

An advertisement for the lease of the brewery and malthouse at London Colney appeared in April 1872. The Colne Brewery was put up for auction in 1873 and 1874 but seems not to have sold. The last mention of the brewery is in August 1877.

MARKYATE

Lion Brewery

James Pearman, born in Aspenden, was a brewer and maltster listed in Cheveralls Lane (later Pickford Road), 'Market Street' in the 1839 edition of *Pigot's Directory*. He died in December 1845 and was buried in the Quaker burial ground in Luton. E. Pearman was one of three brewers listed in Markyate in 1847, presumably James's brother, Edward.

The brewery was next, from at least 1849, in the hands of James and Edward's nephew, Albert Pearman Rowley, born in 1814. In 1857 Rowley acquired what is today 8 Albert Street, off the west side of Markyate High Street. There he opened the Lion Brewery.

Albert Rowley died in April 1884, to be followed at the brewery by his son Alfred William Rowley, born 1850. By around 1882 one of the brewery's mortgagees, Charles Trafford, was living in part of the brewery. In April 1886 the Lion Brewery, the flour mill and the maltings, 'now in the occupation of Mr C. Trafford' were put up for auction 'by order of the mortgagees'.

When the brewery actually closed is not clear, but in November 1891 tenders were invited for the demolition of the brewery and corn mill, with the sale of all the bricks, slates, timbers and interior fittings.

Shemeld, High Street

Robert Shemeld was listed as a brewer, maltster and corn chandler in Markyate in 1838, at what is now 11 High Street. In May 1853 a sale was announced of the farming stock and other property, including brewing equipment, belonging to Mr Robert Shemeld, 'who is leaving Markyate Street'.

It seems likely that it was Shemeld's former premises that were those used by C. Willmott of the Old Brewery, Markyate Street, who was advertising in the *Dunstable Chronicle* in 1858 and is listed as a brewer in 1859 under Flamstead, presumably because he was based in what was then the part of Markyate that came under that village.

PUCKERIDGE

Puckeridge Brewery

Matthew Quinton Fleming, born in Dundalk, Ireland, in around 1807, was a common brewer living in The Cottage, Standon – Puckeridge is in Standon parish – in 1851. Fleming continued to brew in Puckeridge for more than thirty years, dying in 1886, and was followed by one of his sons, Henry or William, until at least September 1888.

The next owner was Charles Chapman, who was advertising his 'genuine home-brewed ales' from the Puckeridge Brewery from July 1889. However, the brewery was soon in financial difficulties, and in September 1892 Charles and Edwin Chapman signed an agreement to sell the brewery to Clement William Clement-Smith. However, the debts Clement-Smith inherited were too heavy for his company to survive, and in February 1894 an official winding-up order was made for the Puckeridge Brewery Company Ltd.

In summer 1894 the brewery was acquired by Henry Runnacles, born in Harwich, Essex, in around 1840. However, just three months later Runnacles fell ill and died at the White Hart in Puckeridge, where he was staying while the dwelling house next to the brewery was repaired.

In December the brewery was acquired by an architect from Forest Gate, North London, named Frederick A. Ashton. He, too, lasted only a short time, and in December 1901, the whole of the brewery plant and stock, including an eight-barrel copper, four

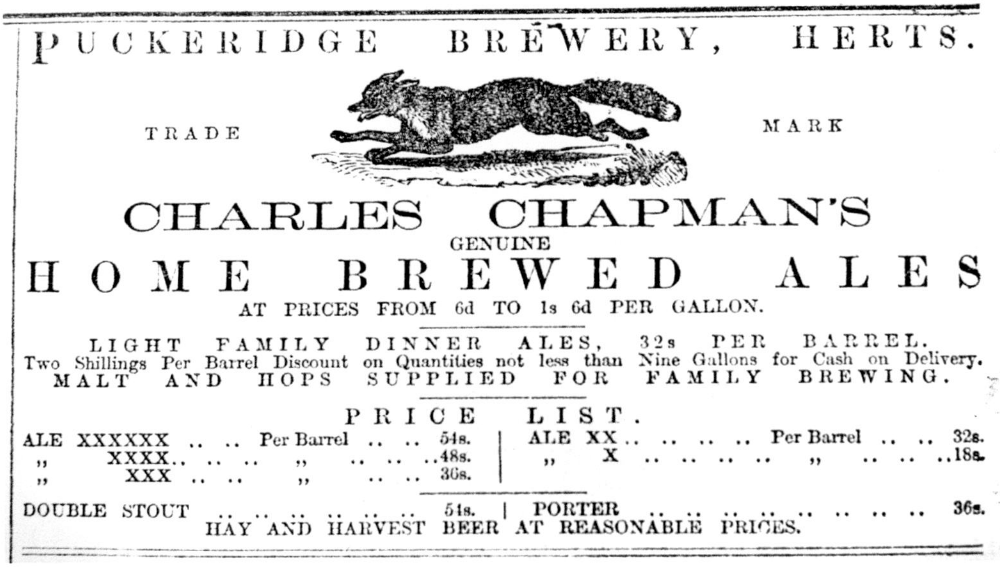

An advertisement for Chapman's Brewery from the *Herts and Essex Observer* in 1889. It features the brewery logo, a running fox, commemorating the Puckeridge Hunt, which has been in existence since 1799. Unusually, the brewer seemed to brew no bitter ales. It also seemed to suffer from X inflation: its XXXXXX was the same price as other brewers' XXXX and would have been around 6.5 to 7.5 per cent alcohol.

fermenting vessels, 317-dozen screw-stopper bottles, twenty-five-dozen dinner ales and stouts and twenty-seven barrels and kilderkins of AK ale and old beer were put up for sale 'under executions from the Sherriff of Hertfordshire'.

REDBOURN

The Redbourn Brewery

The Redbourn Brewery grew out of the Bull Inn. The inn was bought in 1840 by Joseph Liley, forty-nine. Liley stayed at the inn until summer 1849. He then moved on to another inn in the High Street, the White Hart, where in September that year he was promising 'Families supplied with genuine home brewed beer and pale bitter ales'.

The Bull, now owned by Charlotte Lyon-Bowes, the Dowager Lady Glamis, and one of the largest landowners in Hertfordshire, remained closed after Liley's departure. Then in September 1852 an advertisement appeared in the *Hertford Mercury* announcing that the Bull Inn and posting house 'with a Five-quarter Brewery attached (with or without several Public-Houses and Beer-Shops)' was available to be let.

Robert Howie, born in Banff, Scotland, in around 1800, was licensee of the White Horse, next door to the Bull, in September 1853 when he applied to St Albans magistrates for a licence for the Bull, which was granted the following month.

Howie was succeeded at what was now being called the Bull Inn and Brewery in December 1855 by James Kitchener. In April 1858 John Puddephatt, landlord of the Shoulder of Mutton on Market Hill, Luton, announced that he had 'taken the Redbourn Brewery'.

Henry Puddephatt is listed as landlord at the Bull in 1862, and in August that year Andrew Puddephatt handed over the licence of the Bull to Robert Howie.

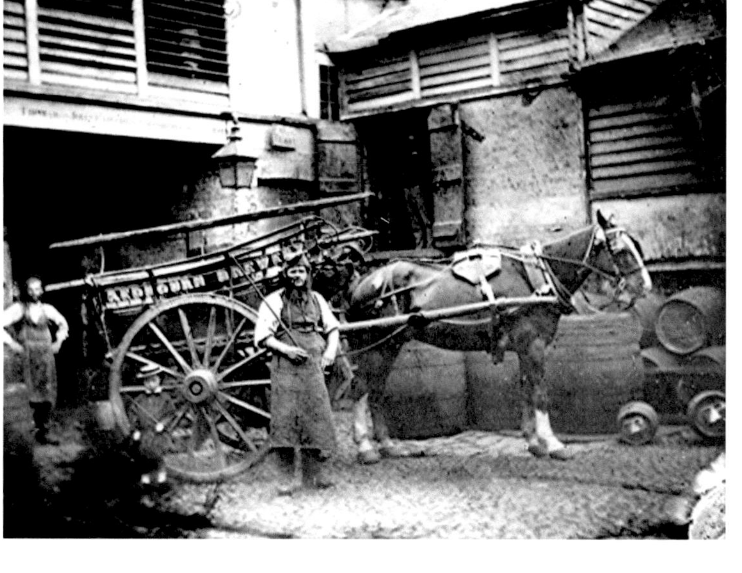

The Redbourn Brewery yard, *c.* 1872. The business was run by the Edwards brothers a this time. The drayman appears to be wearing the traditional brewer's red stocking cap and apron.

By 1866, the Bull Inn and the brewery were being tenanted by Thomas Edwards, born in 1841 in St Albans, and his brother John, born in 1843. The Edwards brothers leased the brewery until 1878, supplying a tied estate of fifteen or so pubs and beerhouses. From March 1879 the brewery and its tied houses were being leased to Benjamin Bennett, the Dunstable maltster, who was already leasing the Harpenden Brewery. In 1887 Bennett acquired the North West Brewery in Dunstable. He continued running the Redbourn Brewery until 1897, when the property was put up for sale.

No buyer came forward for the brewery, and its tied houses were snapped up by other Hertfordshire and Bedfordshire brewers.

RICKMANSWORTH

Salter

The first Salter to turn up in Rickmansworth was Samuel Salter, born in around 1696, who was a maltster in the town from the 1720s, building a malthouse on the way out of town towards Croxley Green. In 1741, when he was forty-six, he appears to have bought the Coach and Horses in Rickmansworth.

In June 1750, Samuel Salter died aged fifty-three. The Coach and Horses and the brewery passed to Samuel's nephew, Stephen Salter, and the following year, 1751, Stephen was assessed for twice the rates his uncle had been paying, charged specifically for the first time against a brewery.

Following the death of Samuel Salter in 1829, his will was found to contain an unusual provision. It stipulated that a cask of beer be left in the brewhouse yard every day for passers-by to help themselves to without payment. Eventually the crowds that the free beer attracted meant it was moved to outside the company's maltings. The *Illustrated London News* carried a description, and a picture of the scene, in its issue of 11 October 1851. But by 1856/7 the rowdiness of the people who gathered to drink the free beer finally forced the brewery to end the tradition.

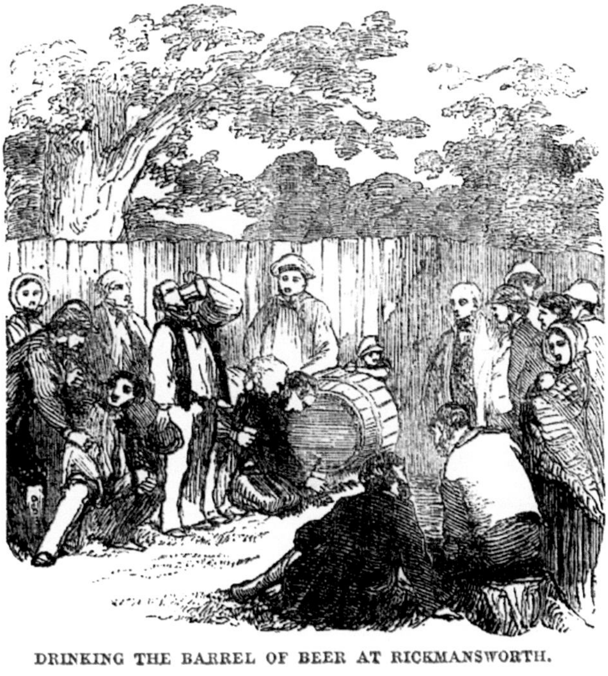

DRINKING THE BARREL OF BEER AT RICKMANSWORTH.

The brewery was probably at the Coach and Horses, but at some point the Salters erected a purpose-built brewery at a house called World's End, some 200 yards north of the inn and on the opposite side of the road. In 1760 Stephen Salter insured a brewhouse, mill house and malthouse at World's End for £350. It had at least sixteen tied houses by 1786.

Stephen Salter died on New Year's Day 1800, aged seventy-four, leaving the brewery business to his son, Samuel.

In June 1802 Samuel's only daughter, named Sarah, after her grandmother, married the Reverend William Capell, half-brother to the Earl of Essex and later chaplain to Queen Victoria.

Part of the River Chess was canalised in 1804, from a new wharf in Rickmansworth more than 500 yards down to below Batchworth Lock on the main canal. The brewery was close to the wharf, and also the main user of the cut, which was often referred to as Salter's Cut. Barges carried casks of beer down the canal to the pubs of Uxbridge, returning with the empties.

Samuel Salter was still apparently running the brewery on his own in 1822, when he was sixty-nine, but by 1827 he had taken a partner, Job Woodman. Samuel died in June 1829. Although Woodman carried on running the concern day to day, ultimate control was placed in two trustees, Thomas Fellowes, a Rickmansworth landowner, and William Capell, presumably Samuel's grandson, Sarah's eldest son, then aged twenty-five.

Thomas Fellowes, the non-family trustee of the brewery, died in January 1858. His son, Herbert William, was evidently already running the business before his father's death, a job he held until at least 1881; he described himself as a brewer of Rickmansworth in 1855.

On 24 December 1889, Salter and Co. was registered as a limited company, the first big brewery in Hertfordshire to turn itself into a limited concern. By 1890 the brewery manager was James Scully, then in his late thirties, who was to continue as manager until his death, aged seventy, just before Christmas 1923. Within weeks

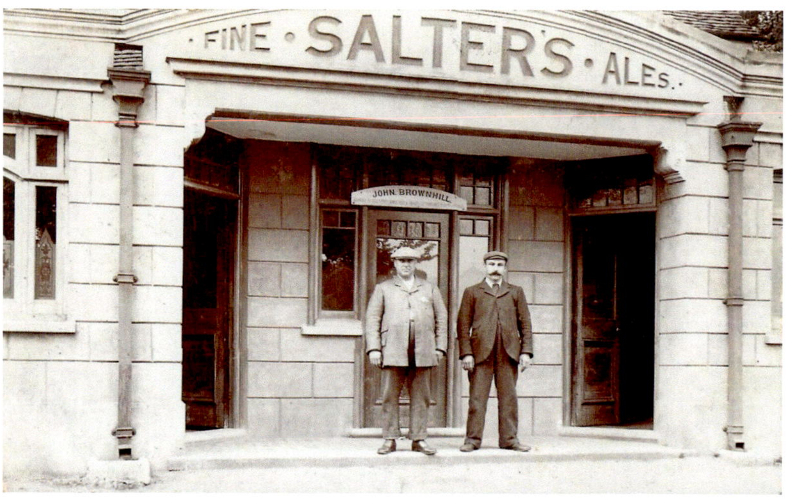

The Green Man at Batchworth Head making it very clear which brewery it was tied to, *c.* 1910. The Salter's sign is today covered by one that says 'Ye Olde Greene Manne'. That is, surely, landlord John Brownhill on the left ...

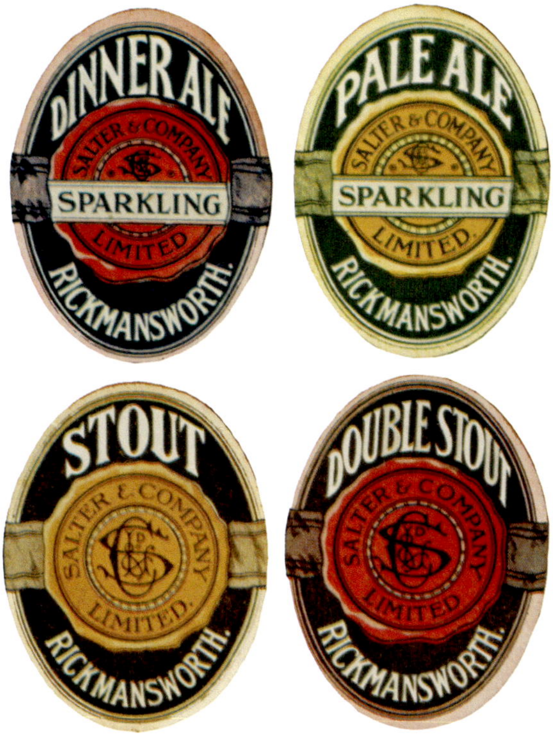

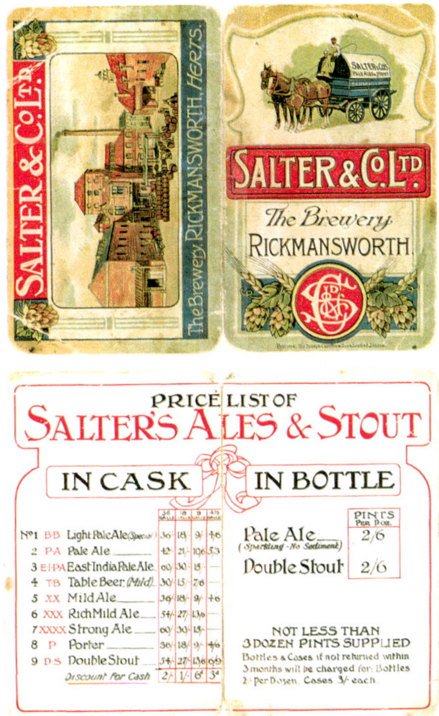

Above left: A set of Salter's labels, two of them promoting the fact that the beers were sediment-free. (Courtesy of Keith Osborne)

Above right: A lovely price list for Salter's beers from the end of the nineteenth century. It details a range of strengths from Light Pale Ale at around 3.5 to 4 per cent abv up to EIPA and XXXX strong ale at 8 to 9.5 per cent. (Courtesy of David Dines)

Salter's directors had arranged the sale of the brewery and all its tied houses to the Cannon Brewery Co. Ltd of St John Street, in the City of London.

Wild, Mill End Brewery

Thomas Wild was a tanner at Mill End in 1839. By 1861 the family had branched out into brewing as well, with that year's census listing Thomas Wild, now sixty-six, as a tanner and farmer and his son, Cornelius Wade Wild, twenty-eight, as a tanner and brewer.

In 1867 the operation was named as 'Thomas Wild and Sons, brewers and tanners'. Thomas died in May 1871, aged seventy-six. In 1881 his younger son, Thomas Wild junior, fifty-seven, was the brewer and farmer, and one of Thomas junior's three sons, Walter, twenty-five, was also listed as a brewer. Thomas died in February 1887 and was followed at the brewery by Walter.

The brewery also made mineral waters, and in 1900 the decision was made to end brewing and concentrate on mineral-water manufacture. The brewery business and six or so tied houses were sold to the Watford brewery Sedgwick's. However, it looks as if the mineral-water business only lasted four years before it collapsed.

ROYSTON

Phillips

The Phillips family, originally of Radwell, near Baldock, owned at least ten breweries at different times, from Suffolk to South Wales and Oxford to Burton-upon-Trent. They were, in the beginning, Quakers, like their fellow North Hertfordshire brewers the Pryors of Baldock and the Lucases of Hitchin. The family's first brewery acquisition came with the purchase by Robert Phillips, a maltster and son of William Phillips of Radwell, in 1725 of premises in Baldock Street, Royston, that had, earlier, at least, in the seventeenth century, been an inn.

By 1759 the brewery was being run by John Phillips, who seems to have been Robert's nephew. He died in 1782, and the brewery passed to his nephew, another John, born in 1762. In December 1805 John II signed an agreement to buy out his rival Royston brewer William Beldam and acquire fifteen more tied houses. The price was £8,000 plus the cost of Beldam's fixtures and fittings.

John Phillips II died in May 1821, aged fifty-nine and unmarried. He had two nephews who shared his name, one the son of his brother Robert, who lived in Royston, and the other, born in 1796, the son of his older brother Joseph, who was running the St Martin's Brewery in Stamford, Lincolnshire. John II left the Royston Brewery to these two cousins, and by 1823 they were running it as John and John Phillips.

Royston John died in June 1826, his share passing to John from Stamford in April 1828. Stamford John, whom we can call John IV, lived in the 'mansion house' facing on to Baldock Street. In 1829 he had the house rebuilt, in pale-yellow gault brick with stone dressings and surrounds, making it one of the finest houses in the district. By 1830 the Baldock Street brewery controlled at least thirty-seven tied houses from Hertford to Cambridge.

Another son of Joseph Phillips of Stamford, Charles, born in 1803, had bought a brewery in Newmarket, Suffolk. In 1865 Charles retired through ill health, and the brewery and nineteen tied houses were put up for sale. John Phillips bought the brewery and six of the pubs, though later four of the former Newmarket pubs were sold.

John Phillips IV died in March 1871. By now the Phillips brewing empire was at its height. The Royston Phillipses had their cousins, Joseph and Thomas Simpson, running

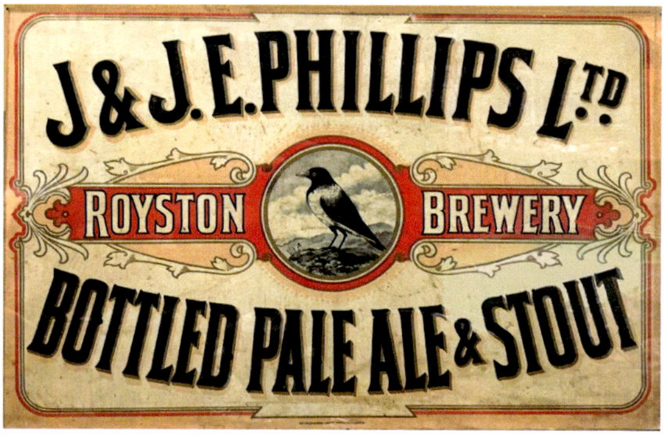

A showcard for Phillips's Brewery showing its 'Royston crow' trademark. 'Royston crow' was the local name for a hooded crow, which is found around Royston, and also gave its name to the local newspaper. (Courtesy of Kevan Witt)

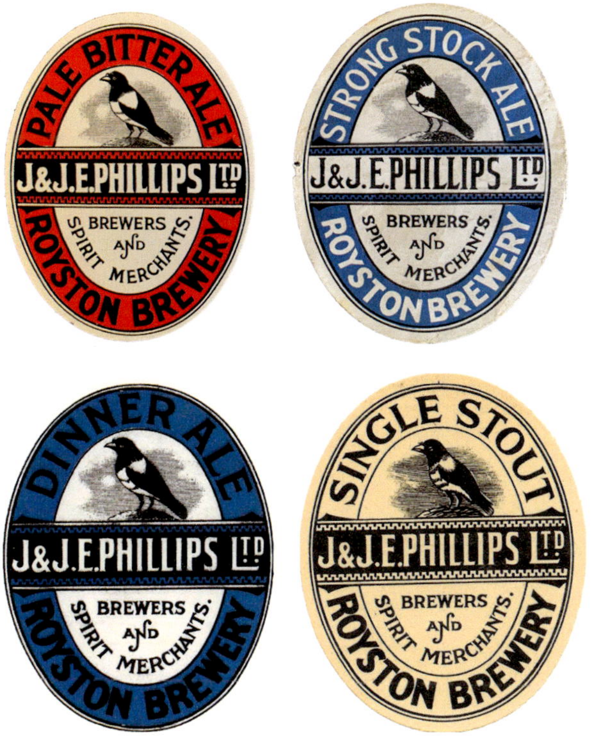

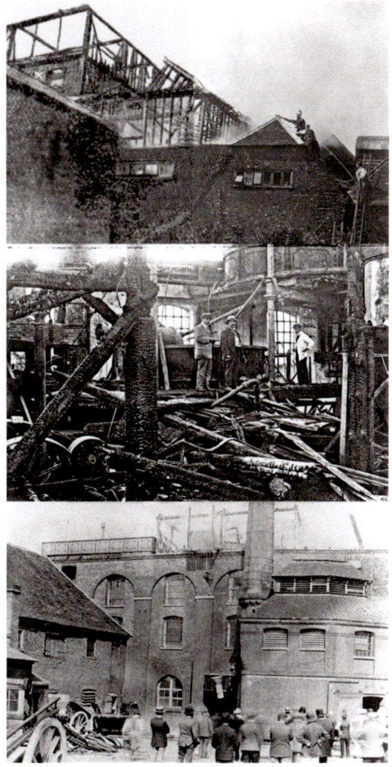

Above left: Four labels from the Royston Brewery.

Above right: Phillips had fire extinguishers in all parts of the brewery, and two hydrants on the premises, with hoses. This did not help early one afternoon in August 1909 when a piece of stone in the malt mill at the top of the brewery sparked as it ground against the mill rollers. Within minutes the light wooden platforms and ventilating shafts topping the main brewery building were well alight. More than 40,000 gallons of water was used to bring the fire under control. Damage was largely restricted to one mash tun destroyed and the new copper and bottling department only slightly affected. Brewing was stopped only temporarily. Top: a fireman plays his hose on the fire. Middle: members, probably, of the Phillips family inspect the damage. Bottom: spectators in the brewery yard. (Photos courtesy of Royston Museum)

the Baldock Brewery down the road. Another cousin was running the St Martin's Brewery in Stamford, while Thomas Phillips, a second cousin to John IV, had built the Phoenix Brewery, Northampton, in 1856. In 1874 Thomas moved to Newport, Monmouthshire, to take over the Dock Road Brewery. His brother, William, who had been running a brewery in Burton-upon-Trent, sold in 1873 to Truman, Hanbury and Buxton in 1873, took over John Higgins's brewery in Oxford. Four years on, in 1878, Thomas's son, Frank, bought Marriott's Midland Brewery in Coventry, which became Phillips and Marriott.

The Royston Brewery acquired Alexander Frederick Tooth's Castle Street brewery in Cambridge in 1899, with ten tied houses, for £29,500.

After the Second World War, the brewery still had some 150 tied houses, but sixty-five of them sold less than a barrel of beer a week each. On 25 November 1949 it was announced that Phillips and its 149 tied houses were to be acquired by J. W. Green of Luton.

In January 1950 the Royston Brewery closed, with each employee being given £7 for every year they had worked for the company.

ST ALBANS

Brewing was probably taking place at St Albans abbey, to supply the monks, from its foundation. A list of abbey servants from 1302 includes Alan 'de Braseria Abbatis' and Walter 'de Braseria Abbatis', 'of the abbey brewery'. There was also Reginald 'de Braseria Conventus', of the convent brewery, who brewed for the nuns and sisters at St Mary de Pré Nunnery.

A trade list for the town compiled around 1360 mentioned a total of eighty-one brewers, though more than half (and probably all) also carried on other trades. All seven of the hostelers listed were also brewers.

By Elizabethan times, only freemen of the borough, or people who had been through an apprenticeship, were lawfully allowed to brew for sale in the town. In 1606 the St Albans Corporation agreed that too much brewing in the borough was driving up the price of fuel, and for that reason the number of brewers should be restricted to four beer brewers and two ale. One of the permitted brewers was John Moseley, who kept the White Hart inn.

Chequer Street Brewery

The first brick of the new concern later known as the St Albans Brewery was laid on 29 May 1775, and the first guile was brewed in January 1777, an 'ancient' board in the brewery storehouse declared, according to a newspaper report in 1863.

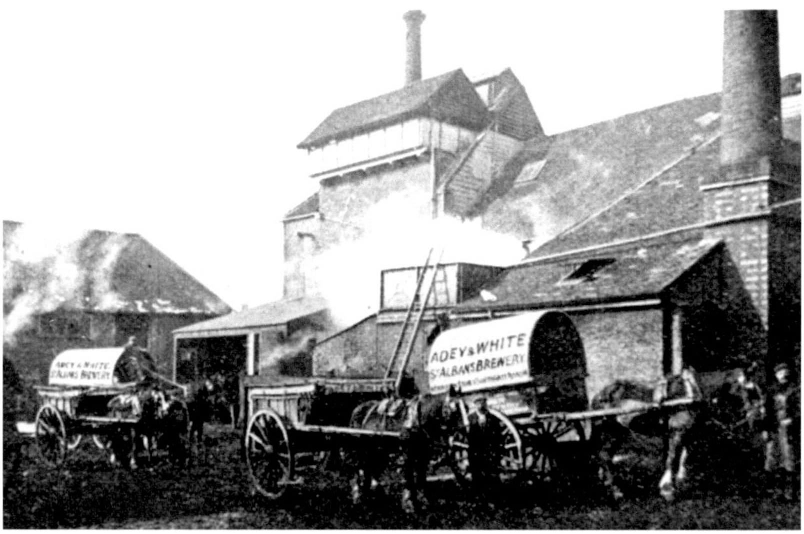

The yard at the Chequer Street brewery, probably photographed between 1868 and 1878.

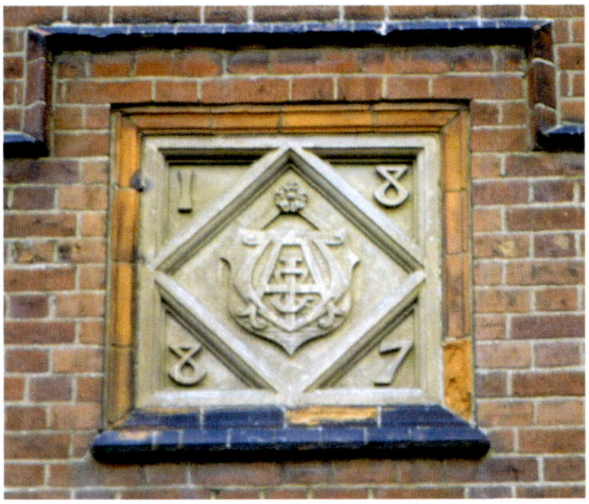

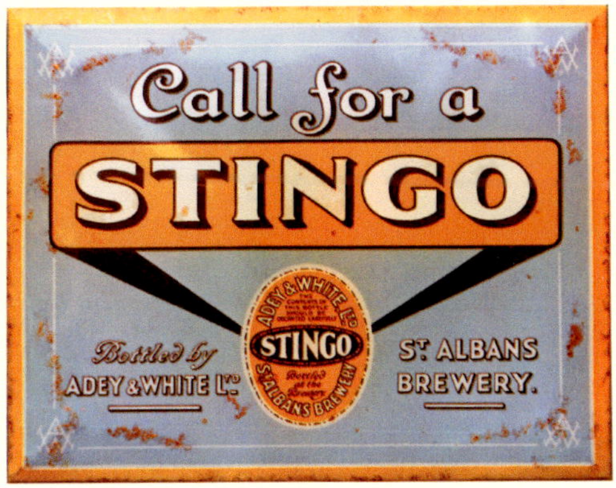

Above left: A plaque on the wall of the former Chequer Street brewery with Adey and White's entwined initials and the date 1887 in a very Arts and Crafts typeface, possibly marking a rebuilding.

Above right: An advertising card for Adey and White's strong Stingo from the 1920s or 1930s.

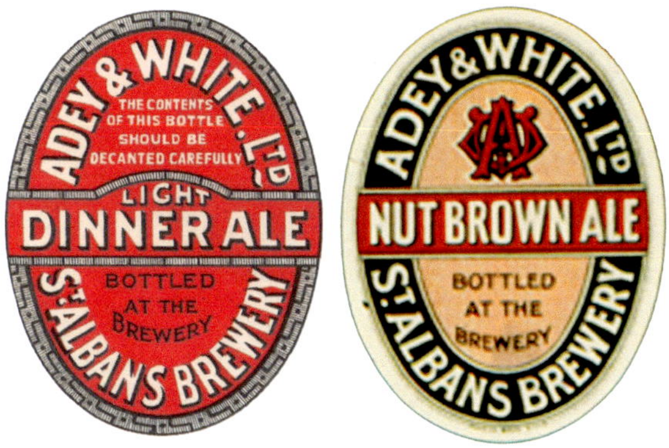

Two Adey and White bottle labels, the light dinner ale clearly a sedimented beer with yeast left in the bottle.

However, the brewery's first owners, the Kinder family, are found as brewers in St Albans from the 1730s. In 1765 Thomas Kinder bought the Bell inn on Chequer Street, St Albans. It could be that Kinder moved into the Bell's own brewery, subsequently rebuilding and enlarging it.

In 1868 the brewery owner was another Thomas Kinder, who retired from business, and the brewery passed to Stephen Vizard Rede Adey and Samuel Monkton White.

Ten years after acquiring the St Albans Brewery, in 1878, Adey and White bought its near neighbour, the St Peter's Brewery, just up the road, for a rental of £1,550 a year. For many years Adey and White brewed at both breweries, almost certainly up to the turn of the century and possibly until a fire in 1902.

White died in 1897 and Adey in 1899. The company was then run by Adey's son, Harold, who was twenty-four when his father died. Harold Adey's position as one of St Albans' leading businessmen was handy during the First World War, since it enabled him to earn exemption from the military call-up.

By the 1930s the brewery was old, and it needed capital expenditure – the wooden fermenting vessels, for example, lacked a copper lining. There were no heirs to carry on the business, for of the two directors Reginald Kent was a bachelor, while Harold Adey only had daughters.

In May 1936 J. W. Green's, the Luton brewer, acquired Adey and White and all its fifty-two surviving pubs and beerhouses and four off-licences.

For many years the site was used as a car park until 1982, when work began to turn it into a shopping centre, The Maltings, together with a new home for St Albans City Library.

Crown Brewery, Verulam Road

In December 1867 G. Cox announced in the *Herts Advertiser* that he had opened the Crown Brewery on Verulam Street (*sic*). By March 1869, Cox had been replaced at the Crown Brewery by William Harvey. In September 1871, Harvey advertised that he was applying to move the licence of the Crown Brewery to 'more eligible premises', his new home in 'Back Street', St Albans, Back Street being another name for French Row.

By January 1874 his business was in liquidation, and his estate included a 'valuable small brewery newly erected on the tower principle' and mashing four quarters at a time, and the lease of the Old Crown Brewery Tap was put up for auction at St Albans Corn Exchange.

It seems to have failed to find a buyer then and at three more auctions. The Crown Tap was renamed the Golden Harp by 1877.

T. W. Kent, Holywell Brewery

When the Holywell Brewery, owned by the Kent family, merged with its rival Adey and White in 1918, it was claimed that the brewery had been founded in 1826. There seems no evidence of this: the earliest the Kents are mentioned as brewers appears to be 1838, when Thomas Weedon Kent, then thirty-two, was brewing in Chequer Street, possibly at the Bat and Ball beerhouse.

Kent continued to be listed as a brewer in Chequer Street until 1860. By 1861, however, he had moved to Holywell Hill, next door to the former Woolpack Inn, an old coaching inn that had brewed its own beer. The Woolpack was bought in December 1852 by William Marks, the owner of the next-door inn, the Peahen, and vanished as a separate

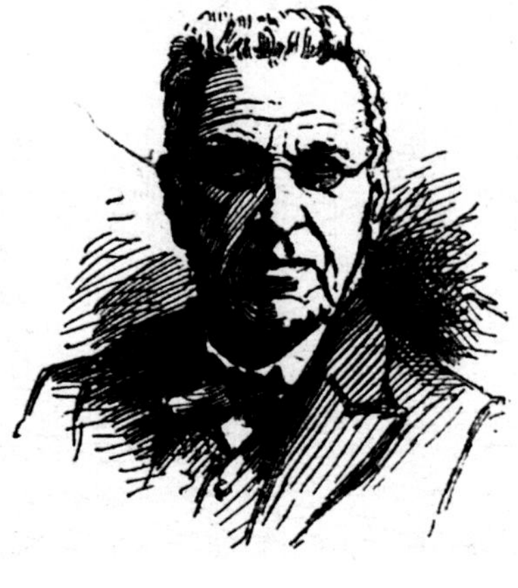

The stern visage of Thomas Weedon Kent.

establishment after 1856. It may be that Thomas Weedon Kent moved his brewing operations from Chequer Street to the former brewery at the Woolpack on Holywell Hill.

In 1900 Thomas Weedon Kent died, aged ninety-four, to be succeeded as head of the business by his son, Thomas Kent III, born in 1839. Thomas Kent III died in 1917, aged seventy-eight, by which time his son, Reginald, was running the brewery. The following year, in October 1918, the little firm merged with its much larger rival, Adey and White, and the Holywell Brewery was closed.

Kingsbury Brewery

A newspaper in February 1722 reported a fire at Alderman Francis Carter's brewhouse in St Albans which caused 'considerable damage'. This appears to be the first mention of the concern. The brewery, if it was on the same site as the brewery owned by Carter's descendants, was on the east side of Fishpool Street.

Carter had a daughter called Dorothy. She married a man named William Nicoll, or Nicholl, from Hatfield in 1720, and they had a son called Francis Carter Niccoll, born in around 1723. Carter died in 1743, and his grandson eventually ran the brewery.

In 1750, when Francis Carter Niccoll was mayor of St Albans, he married Anne Searancke, whose brother, John, owned the big brewery in Hatfield. Their son, Francis Nicholl junior, born in 1752, took the surname Searancke in order to inherit the Hatfield Brewery under the terms of his uncle's will, which he did in 1789.

He was already the owner of the Fishpool Street brewery following the death of his father in 1782, aged fifty-eight. In 1815, he sold his share of the Hatfield Brewery to his partner there, Joseph Bigg.

His son, Francis Joseph Searancke, born in 1810, moved the brewery, and his home, in 1837 to a newly erected brewery and house in Verulam Road, St Albans, near Kingsbury Lodge. Francis Joseph Searancke retired in 1868, and the brewery and its tied houses were leased to a local landowner, John Lloyd, aged twenty-eight, of Astwick Manor.

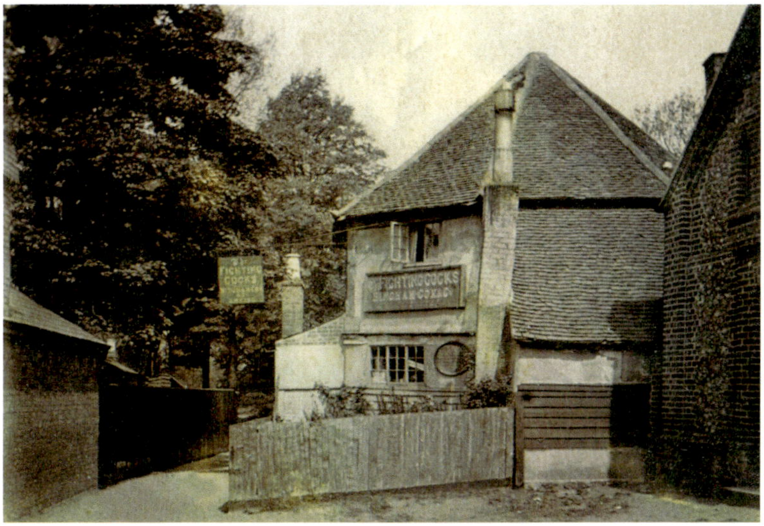

The Fighting Cocks – NOT the 'oldest pub in England', and not even the oldest pub in St Albans – bearing the name of Bingham-Cox & Co., sometime between 1889 and 1894.

In December 1888, Francis Joseph Searancke died, and his trustees put the Kingsbury Brewery up for sale. Lloyd retired, and on 15 July 1889 the brewery and the thirty-two pubs, inns and beerhouses owned by the Searancke family were auctioned at the Mart in London. A week later Lloyd's own twenty-six pubs and beerhouses were auctioned at the same place.

Someone on the fringe of the trade, William Bingham-Cox, editor of the *Licensed Victuallers' Gazette,* bought the brewery and its tied houses for £24,100. At the second auction, of the Lloyd-owned pubs, Bingham-Cox only managed thirteen successful bids, and six other brewers shared the rest.

Bingham-Cox, born in Gloucestershire in 1841, had started training as a barrister, but left and went into journalism. He raised the circulation of the *Licensed Victuallers' Gazette* from 1,500 copies a week to over 50,000, and he was reckoned to be worth more than £100,000, equal to perhaps £12 million today.

In July 1894 Bingham-Cox put the brewery, with fifty-two tied houses, up for auction. No sale was agreed, but a few months later, at the end of 1894, Bingham-Cox accepted just £50,000 for the whole concern from a newly formed company, the Kingsbury (St Albans) Brewery Co. Ltd. The Kingsbury Brewery, with all its tied houses, was then taken over by Benskin's of Watford in 1897. The brewery seems to have continued in operation until 1900, when it finally closed.

St Peter's Brewery

The St Peter's Brewery in St Peter's Street was founded by Henry Parsons, and was in existence by 1832, when Parsons was only twenty-five.

From January 1870 the business was run by Henry's younger son, Francis, then aged twenty-seven. In April 1878 he put the St Peter's Brewery, then producing 6,000 barrels a year, and its thirty tied houses up for sale. It appears not to have sold, and instead the lease of the brewery was acquired by its rival a few hundred yards away, Adey and White, who carried on brewing at the St Peter's Brewery.

Parsons leased fourteen of the brewery's tied houses to Adey and White. Other pubs in the estate were disposed of to a string of different brewers.

SANDON

Hyde Hall Brewery

When George Clark first took up the lease of Hyde Hall Farm, Sandon, in 1830 it included a brewery with a hogshead brewing copper, a deal cooler, an iron-bound mash tun and three iron-bound brewing tubs.

By 1862 George's second son, Henry Boon Clark, born in 1842, looks to have turned the farm brewery into a commercial operation. Henry brewed at Hyde Hall until at least 1864 but by 1866 he was farming and brewing in Whittlesford, Cambridgeshire. (He went on to have a career that took him around the country, ending up running a brewery in Wakefield Yorkshire.) In 1871 his older brother, George junior, born in 1830, was running the farm and the brewery at Hyde Hall.

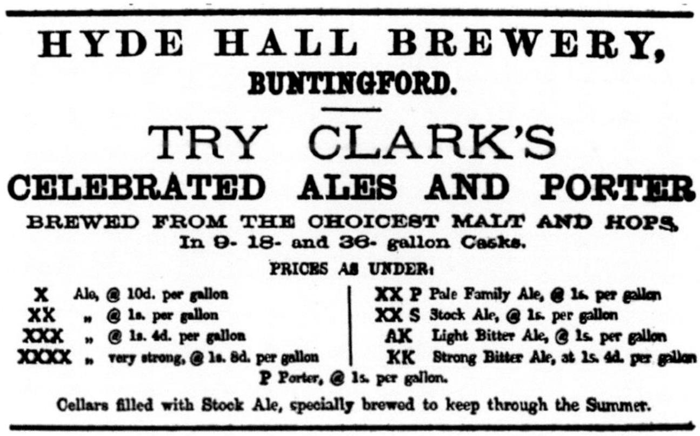

An advert for the Hyde Hall Brewery from the *Royston Crow*, February 1878.

Another brother, Joseph Porter Clark, was running the brewery by 1884. He was only the manager, however, and the brewery was owned by William Robert Baker of Bayfordbury, a wealthy Hertfordshire landowner. Clark and Baker eventually fell out in or just before January 1888 and the Hyde Hall Brewery continued under Baker.

In July 1893 the brewery, along with four tied beerhouses in Buntingford and Barkway, was put up for sale. It failed to sell, and Baker continued running the business until May 1898, using a manager. Baker then sold the brewery to Fordham's of Ashwell for a total of £695 6*s*.

Fordham's continued brewing at Hyde Hall for a year before closing it.

SAWBRIDGEWORTH

George and Henry Rochester

The George IV in Knight Street probably opened in 1822. After George Rochester began leasing the George IV in 1852, he rapidly began brewing there. He was still brewing at the pub in 1861, but by 1866 it had become a Christie's of Hoddesdon tied house. However, George Rochester found himself new premises in Knight Street, and carried on brewing.

George died in 1897, and the brewery business was continued by his son, Henry, until at least 1906, but had evidently stopped by 1908.

William IV

James Rochester was certainly a beer retailer by 1833, when he was forty-eight, almost certainly at the William IV, Fair Green. By 1841, and probably from the start, he was also a brewer at Fair Green.

James looks to have died in 1846, and in 1850 the beerhouse and its brewery were being carried on by his widow, Elizabeth. In 1851 she was a brewer and retailer of beer, living with her son, George, thirty, his wife, Sarah, and their new-born son, Henry.

The following year, however, the William IV and its brewery were taken on by George's older brother, Samuel, born in 1807. Samuel brewed at the William IV until his death in 1874. His widow, Susannah, continued running the brewery and beerhouse.

In September 1886, the King William IV was bought by Thomas Chaplin, owner of a brewery in Harlow, Essex, for £600 and brewing at the William IV stopped.

SHENLEY

Wheatsheaf Brewery

A public house called the Wheatsheaf with its own brewery had sprung out of Wheatsheaf Farm, Well End, by the mid-1820s.

In 1850 Henry Bennet Smith was listed as a brewer at Shenley, and the census the next year shows him as a farmer and brewer. Meanwhile, Benjamin Tuck was apparently living at the Wheatsheaf, his occupation, too, given as brewer.

Tuck was still brewing at the Wheatsheaf until his death in 1867. In 1870 the pub was being run by his widow, Margaret, while the brewery was being run by the farmer at Wheatsheaf Farm, Thomas Alexander Williamson. The 1871 census shows Williamson, twenty-five, and his brother, Frederick, twenty-one, both described as brewers at Wheatsheaf Farm.

A map of 1877 shows a building labelled 'Wellend Brewery' alongside the Wheatsheaf pub, but there is no other evidence of the brewery still being in operation at this date.

STANDON

Standon Brewery

William Garratt, then twenty-five, and his brother-in-law, Andrew James Waller, twenty-seven, were living at Mill House, Mill End, Standon, in 1859 and running the Standon Brewery.

Garrett left shortly after 1865 and Waller, born in Welwyn in 1831, continued at Mill House as a farmer, brewer and miller. He looks to have been employing a professional brewer, Joseph Rogers, born in Princes Risborough.

Waller is listed in 1882 as running a brewery at the farm, but the brewery was gone by 1886.

STANSTEAD ABBOTTS

John Cooper

John Cooper, born in Geddington, Northamptonshire, in around 1781, was listed as a brewer in 1838. In 1813 the marriage of Mr John Cooper of Stanstead distil-house to Miss Perkins of Lambeth was recorded – a distil-house is an old name for a distillery.

In 1851 the census lists Cooper as a brewer while his son was a brewer's clerk. John died that year, and his son was listed as a brewer in 1855 but was gone by 1859.

Cornelius Perkins

Cornelius William Perkins was born in Stanstead in 1816. He was a blacksmith and a farmer of 35 acres in 1851. Perkins had added brewing to his trades by 1862, when he was described as a brewer on the birth of his son. By 1869 he was the owner of the Five Horseshoes pub in the village. In 1871 his sole profession was 'brewer', and in 1891 he was a brewer employing three men. He died in 1899, aged eighty-three.

STAPLEFORD

Woodhall Arms

The Woodhall Arms pub in Stapleford was in existence in 1837. It was owned by Abel Smith, the wealthy MP for Hertfordshire and owner of Woodhall Park, between Stapleford and Watton-at-Stone. The landlord at the Woodhall Arms in 1846 was Joseph Neale. In 1850 his son, Charles, born in 1824 in Great Amwell, was described as a brewer of Stapleford. In the following year's census, Charles was recorded at the Woodhall Arms as a brewer employing two men.

The landlord at the inn in 1878, George Ayliffe, continued to be listed as a farmer and brewer at the inn until 1891. By 1894 he was gone, and by 1902 the inn was a McMullen's tied house.

STEVENAGE

Buckingham Palace

John Buckingham, born in Preston, near Hitchin, in around 1802, bought the property in Middle Row that was to become the Buckingham Palace in 1852. He soon installed a brewing plant at the premises, and almost as quickly he was in financial trouble: by November 1853 he was mortgaging his brewing plant and other goods to raise cash.

Brewing seems to have stopped at the Buckingham Palace soon after, but Buckingham carried on running the beerhouse, and baking, until he died in 1876 and the pub passed to his son, Charles. At some point before 1879 the Buckingham Palace was mortgaged to the Baldock brewer Simpson's, and in 1881 Simpson's foreclosed.

Griffin Cant

Griffin Cant, from Great Staughton, Huntingdonshire, was a brewer and corn, seed and coal merchant in Stevenage in 1845, very probably in the premises previously run by Hicks and Richardson (see next page). That year, however, he became bankrupt, and in 1846 'by order of the assignees of Mr Griffin Cant', the contents of the 'very snug' eight-quarter Stevenage Brewery, 'near to the Swan Inn', were put up for sale.

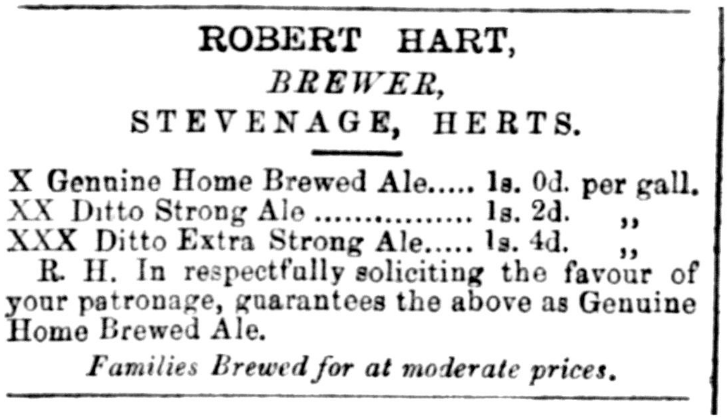

ROBERT HART,
BREWER,
STEVENAGE, HERTS.

X Genuine Home Brewed Ale..... 1s. 0d. per gall.
XX Ditto Strong Ale .... 1s. 2d. ,,
XXX Ditto Extra Strong Ale..... 1s. 4d. ,,

R. H. In respectfully soliciting the favour of your patronage, guarantees the above as Genuine Home Brewed Ale.

Families Brewed for at moderate prices.

An advertisement for Robert Hart's Brewery, Stevenage, in the *Herts Express*, January 1867. The brewery produced just three different types of ale.

Robert Hart

Robert Hart was advertising himself in January 1867 as a brewer of 'Genuine Home Brewed Ale'. He appears in Albert Street as a brewer in 1871 but disappears after that.

Hicks and Richardson

William Hicks and Samuel Richardson were recorded as brewers and maltsters in 1839, their home immediately to the south of the Swan inn, according to the 1841 census. Their maltings was probably the one next door to Alleyne's school. In September 1843 the partnership between William Hicks and Samuel Richardson, brewers of Stevenage, was dissolved.

The Mouldens

It looks as if George Moulden started in business as a brewer in 1859, when he was listed as a carpenter and brewer four doors down from the Unicorn pub in the High Street. He was listed as a brewer in 1882 but retired soon after.

Other members of the Moulden family also brewed: in 1838, William Moulden was a brewer and cooper; Mary Moulden, born in around 1811, was a brewer in 1841; and John Moulden, a cooper employing one man and living in 1851, probably, at what was later 115 High Street, which became a little unnamed beerhouse at some time, was a brewer as well in 1854.

John Moulden continues to be listed as a brewer and cooper in directories until 1870.

Swan Inn

The Swan in the High Street, today The Grange, was the principal inn in Stevenage, with its first known mention in 1530. The first mention of a brewhouse at the Swan seems to have been in 1830, when it was put up for sale.

In July 1847, the seven-barrel brewing copper at the Swan together with two coolers, a five-quarter mash tun and a mass of building materials were auctioned off at the inn. Soon after, the last landlord at the Swan, Thomas Cass, closed and sold it.

TRING

Akeman Street Brewery

William Olney, brewer of Tring, was mentioned in February 1837. He was also a canvas manufacturer, based in Akeman Street. William looks to have died in 1852, and his sister, Sarah, then aged seventy-four, was listed as a brewer in 1854.

Sarah died late in 1858 or early in 1859, and her home in Akeman Street plus the attached six-quarter brewery and a canvas weaving shop to the rear were put up for auction.

No buyer seems to have come forward, and it looks as if the brewery remained empty for around twelve years. Then, some time between 1871 and summer 1873,

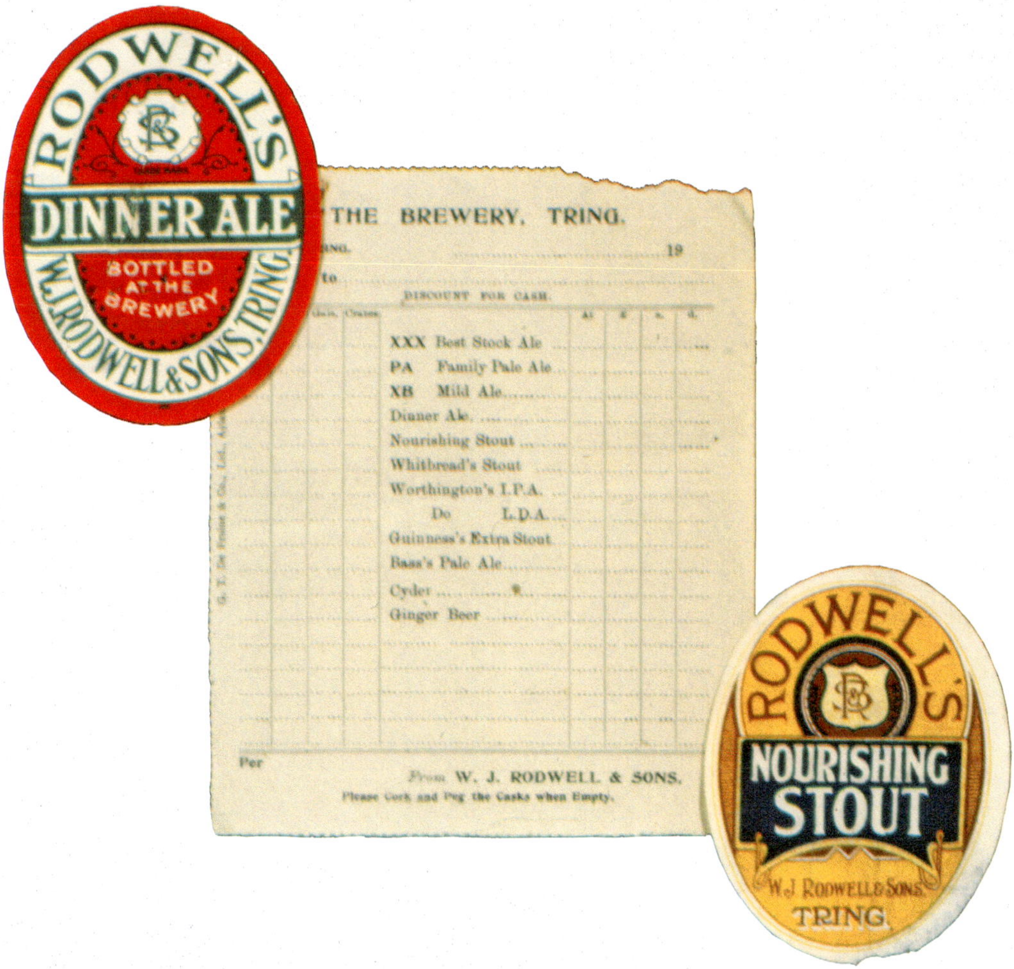

Two labels from Rodwell's Brewery in Tring, plus a receipt that shows the firm was also selling beer from four of the big brewers. (Courtesy of Keith Osborne)

John Batchelor moved his brewery, founded by 1854, from the village of Hastoe, a mile south of Tring, to the former Olney Brewery.

Shortly before 1894, Batchelor, born in around 1830, acquired a partner in the brewery business, his son-in-law, William Jesse Rodwell, who was running the brewery on his own by 1898.

In June 1923 the firm announced that it was giving up brewing to concentrate on mineral-water manufacture.

Green Man

The Green Man in the High Street was kept by the Tomkins, or Tompkins, family for many years. In the 1830s it was being run by Joseph Tomkins, born in 1800, who had married Jane Meager in 1829. After Tomkins's death in 1839, Jane married John Philby (or Philbey), plumber of Tring, who was a brewer and innkeeper at the Green Man in 1845.

Philby died within a few years, and by 1851 the inn was being run by the now twice-widowed Jane, forty-seven. She died in 1864, and the inn and brewery were taken over by her nephew, John Meager. He died in January 1872, aged just thirty-three.

By September 1872, John Woodman was the innkeeper and brewer at the Green Man. Woodman, born in 1837, the son of a Tring farmer, continued brewing at the Green Man for the next thirty years. He died in January 1903, and soon after the inn closed and was demolished by Lord Rothschild, the land it stood on incorporated into Tring Park.

Liddington, Manor Brewery

A bottle label issued for Seabrook Liddington of the Manor Brewery on the corner of Tring High Street and Lower Dunsley said that the business was founded in 1837.

By 1861 Seabrook's eldest son, James, was living at the brewery. He ran the brewery until his death in April 1885, aged fifty-five. The Manor Brewery continued under James's widow, Rebecca, until at least 1894, when she was fifty-seven. Shortly afterwards, in 1895/6, the brewery was demolished by the Rothschilds and the site incorporated into their garden.

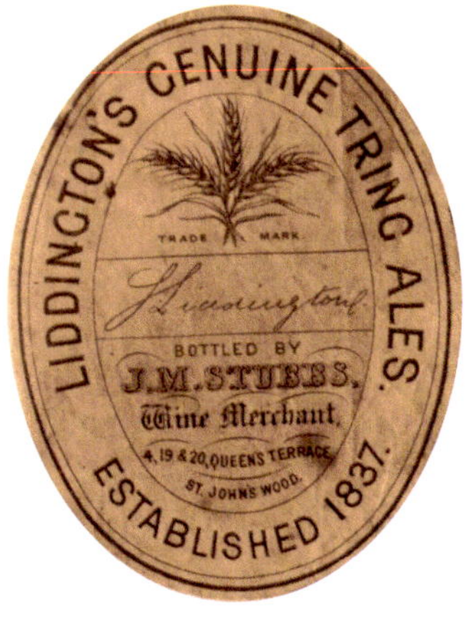

A label for Liddington's Tring Ales as bottled by John Morris Stubbs, grocer and wine merchant, sometime between 1862 and 1885. (Courtesy of Grover Liddington)

Rose and Crown

The Rose and Crown was in existence in 1620. The landlord in the 1820s and 1830s was Timothy Northwood, and he is listed up to 1838 as a brewer and maltster. After his death the inn was continued by his widow, Sarah, to at least 1851.

In 1854 James Shearman was landlord at the Rose and Crown, and also 'ale and porter brewer'. Shearman (or Sheerman), born in 1812, was still a brewer and victualler at the Rose and Crown in 1861. Then on 1 February 1864 the brewery plant at the inn was put up for auction.

Tring Brewery

The Tring Brewery, in the High Street, was supposedly founded in around 1800, and its earliest known owner is said to be Thomas Amsden, who was also a straw-plait dealer, and who is mentioned as a brewer in the town in 1822. However, given the persistence of Thomas Amsden as a brewer through into the 1830s, it seems likely his was not the Tring Brewery, but another concern.

The Tring Brewery was in the hands of James Field by some time in the late 1820s, and Field built up an estate of nine tied houses.

In around 1829 Field retired. He leased the brewery to John Brown, son of a farmer, born in Okeford Fitzpaine, Dorset, on 31 December 1795. Brown added more pubs, including two, the Britannia, Western Road, and the King's Arms in King Street, built for the navvies working on the London to Birmingham railway in the mid-1830s.

By 1881, when John Brown was eighty-five, the brewery was being run by his eldest son, Herbert. He had trained as a brewer at Mew's Brewery in Newport, Isle of Wight.

John Brown died in 1890. Six years later Herbert Brown died, aged fifty-four. The brewery property, including nine tied houses, passed to his younger brother, Frederick William. Frederick ran the business for only a couple of years before it was disposed of to Locke and Smith of Berkhamsted.

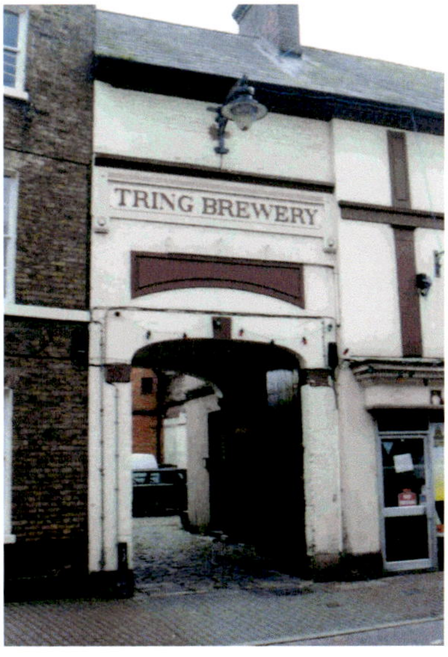

The entrance to the Tring Brewery on Tring High Street, *c*. 1985.

WALKERN

Wright's

Wright's claimed to have begun business in 1799, when Samuel Wright started a maltings operation in Walkern. His – probably – great-grandson, another Samuel Wright, then twenty-three, was a maltster and farmer in Walkern in 1850. In the mid-1850s Samuel turned his farm brewing into a commercial operation. In 1855 he was advertising in the *Luton Times and Advertiser*: 'Pure ales and stout brewed with water from our own artesian well. Also aerated waters of the finest quality … Samuel Wright & Co, Brewers and Maltsters, Victoria Brewery and Mineral Water Works'.

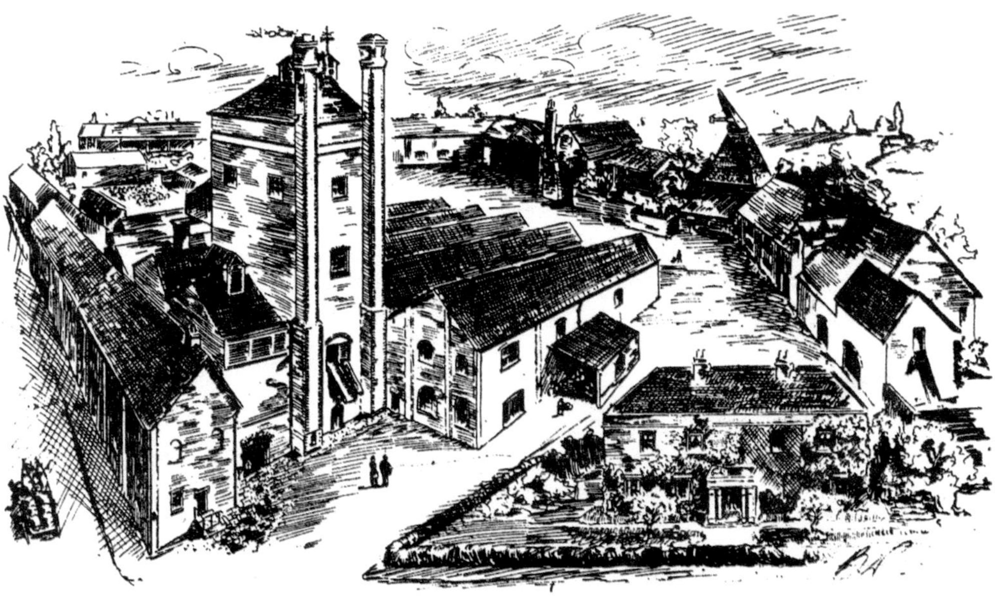

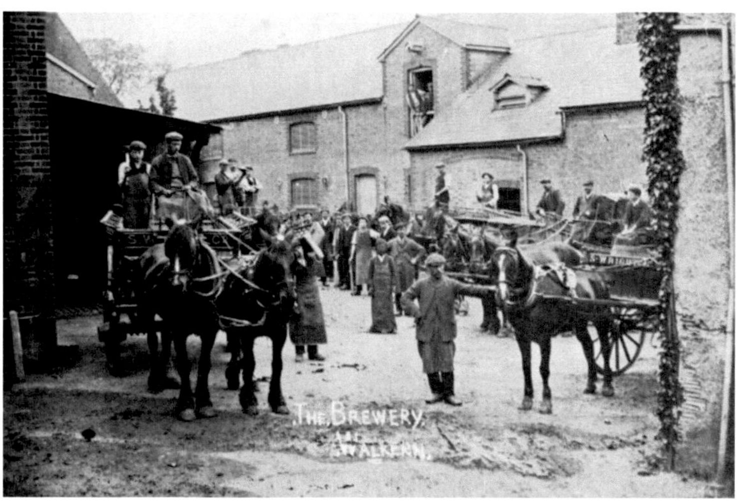

Above: Wright's Brewery in Walkern as depicted in the *St James's Review* in 1907.

Left: Wright's Brewery yard, *c.* 1900. All the workers have come out to have their picture taken. (Courtesy of the Walkern Society)

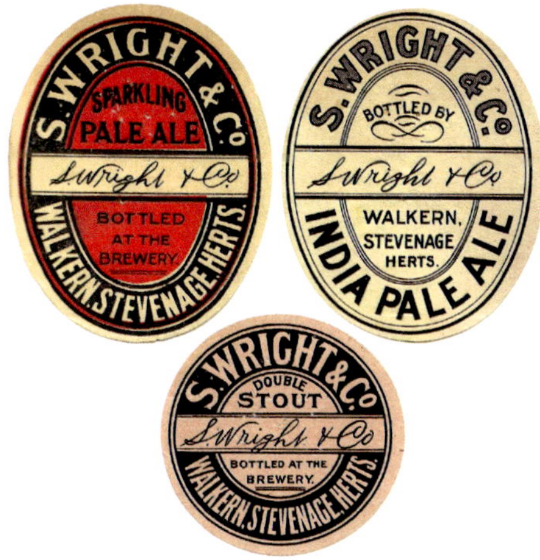

Three labels for Wright's Brewery, the double stout probably dating from *c.* 1870, the other two from *c.* 1900. (Courtesy of Keith Osborne)

By 1871 Wright was employing thirteen men and boys, including a full-time brewery manager, James Holland, twenty-six, from Atherstone, Warwickshire. Two years later, in 1873, the brewery was badly damaged in a fire. Wright erected a brick-built tower brewery on the site, designed by Arthur Kinder, one of the best-known brewery architects of the day.

Samuel Wright died the following year, aged just forty-eight. His two sons, Samuel Eustace Wright and Herbert Wortham Wright, were just fourteen and four. James Holland stayed on as brewery manager, and the business continued as 'Executors of Samuel Wright'.

In around July 1886 Wright's bought Adams & Co.'s brewery in Bridge Street, Nuneaton, Warwickshire, which was turned into an agency for Wright's ales and stout. The Nuneaton brewery buildings were sold off in January 1890.

Wright's bought J. P. Clark's ale and porter stores in Buntingford in March 1889, from the Hyde Hall Brewery, Sandon.

In 1917, Wright's found its cider supplies had dried up. It started making its own, and Wright's cider proved a success. In 1924, Samuel and Herbert Wright decided to stop brewing beer and concentrate on cider and soft drinks. Their sixteen tied houses were sold to Simpson's of Baldock.

Cider making stopped in 1956. Soft drinks continued to be made at what became the Victoria Works until 1985.

Yew Tree

The Yew Tree was first mentioned in 1737, and was possibly brewing from the start.

Henry Elliott was a bricklayer and publican at the Yew Tree in 1851. The 1861 census found him running the Yew Tree brewery and public house and employing six men and four boys.

Later that decade the Yew Tree was sold to William Baker's Hope Brewery in Hertford, and the two-and-a-half-quarter brewery plant, including four-and-a-half-barrel copper, cooler, three squares, horse-powered malt mill and 'excellent' refrigerator, were sold off in 1867. In 1871 the pub was being kept by Elliott's widow, Eliza.

WALTHAM CROSS

Hope Brewery, Marsh Lane

William Walker was brewing in Marsh Lane in 1839. He was a brewer and beer retailer in 1854. By January 1856 Walker had gone, apparently replaced by George Collier of the Hope Brewery, Marsh Lane. Collier was still at the Hope Brewery in 1864, but apparently gone by 1866.

WARE

Bridge Brewery

David Lambert was a retail brewer in 1846 when he died and his brewery and beerhouse at the corner of the High Street and Star Lane were put up for auction. He was followed by George Powter, brewing at Bridge Foot between at least 1851 and 1861.

By 1864 Jordan Sell was at the Bridge Brewery. In 1886, Sell transferred the business to John Jeffrey, or Jeffreys, who ran the brewery until at least 1898. By 1902 Jeffrey was a beer retailer only.

Cannon Brewery, Princes Street

Thomas Cox, born in Bethnal Green, had opened the Cannon Brewery by 1848. Cox continued at the brewery until November 1863, when he sold the brewery to McMullen and Sons of Hertford. McMullen's ran the Cannon Brewery until 1891, when it opened a new brewery in Hertford.

Church Gate Brewery, West Street

Robert Page began brewing in around 1832. In 1845 he bought premises in Crib Street which became the Albion beerhouse. By 1848 the operation was called the Church Gate Brewery.

By 1859 the business was being run by Page's son Charles. In 1866 the Church Gate Brewery and eight beerhouses were put up for auction 'by order of Mr Charles Page', who was retiring. However, Page withdrew the brewery from auction, and a short while later, aged thirty-eight, shot himself in the head.

The three-quarter brewery was put up for auction again in January 1867. The tied houses were sold but the brewery did not find a taker.

New Road Brewery

Some time after 1874 Henry Barker built a two-storey brewery in the back garden of his house in New Road, and by 1878 he was in business as a brewer. The house itself became a beerhouse, the Two Brewers.

By 1898 his son, Silas, born in around 1856, was in charge. Silas ran the brewery until July 1929. It was then acquired by Christie and Lucas, a firm running the off-licences that had belonged to Christie's of Hoddesdon.

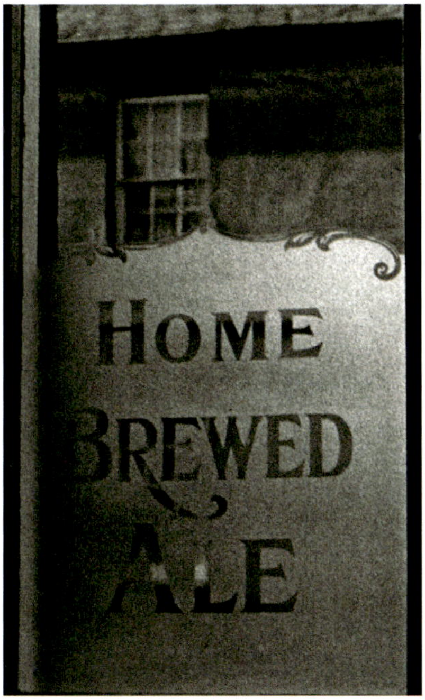

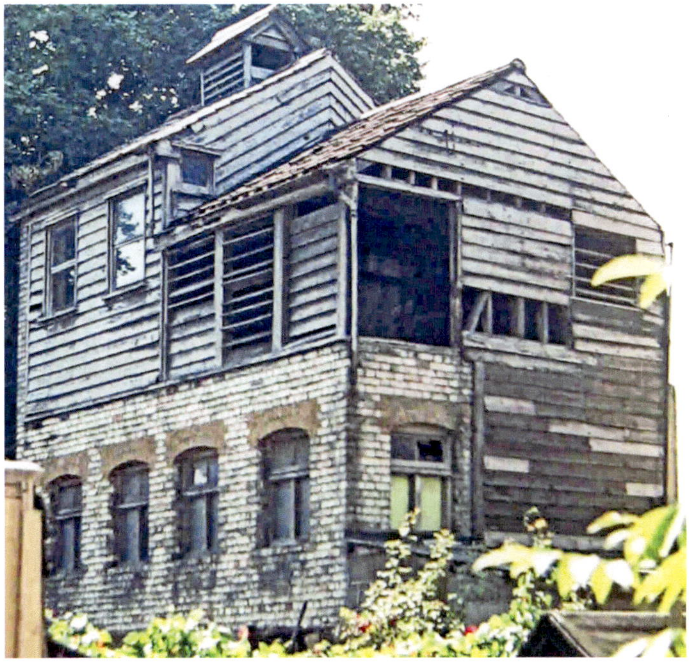

Above left: A window (photograph reversed) at the former Two Brewers, the tap of Barker's Brewery in Ware.

Above right: The former Barker's Brewery in New Road, Ware, 1973. (Courtesy of the Brewery History Society)

Christie and Lucas ran the brewery for barely a year. It remained empty until 1937, when it was acquired by Herbert Lynn under the name of the Ware Brewery Company. However, Lynn moved to the Star Brewery premises in 1938.

Saracen's Head

The earliest definite mention of a brewery at the Saracen's Head seems to be a listing for Daniel Brown in 1839. Brown died early in 1864, aged around sixty-five, and the lease of the inn came to Silas Chuck, a member of a Ware malting family, who is listed there in 1867.

Chuck ran the inn and brewery until 1872, when it was taken over by Isaac Everitt, who ran it alongside the Star Brewery. In November 1874, Everitt's business was acquired by McMullen's of Hertford, and brewing at the Saracen's Head ceased.

Star Brewery, Watton Road

The Star Brewery was founded by Caleb Hitch junior in around 1862. He built up a small estate of three or so tied houses, but ceased business in around March 1869. The lease was acquired by Isaac Everitt. After Everitt's business was acquired by McMullen's in 1874, the brewery was leased by William Wickham.

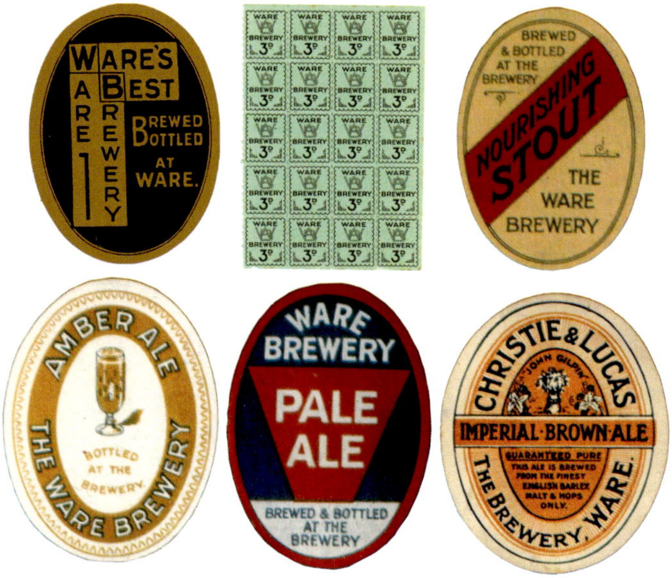

A selection of Ware Brewery labels, one from Christie and Lucas, and a set of Ware Brewery savings stamps. (Courtesy of Keith Osborne)

After Wickham stopped brewing in 1911, the Star Brewery remained empty until 1938, when Herbert Lynn and the Ware Brewery Company moved in. The Ware Brewery Company closed in 1951.

William Wickham

John Long was a farmer and brother-in-law of Edward Wickham of the Mill Bridge Brewery in Hertford. In 1850 he was listed as a commercial brewer, running the High Street Brewery.

Long handed over the High Street Brewery in July 1865 to his nephew William Wickham, elder son of Edward Wickham.

In June 1875 William announced he was now leasing the Star Brewery and running it alongside the High Street Brewery.

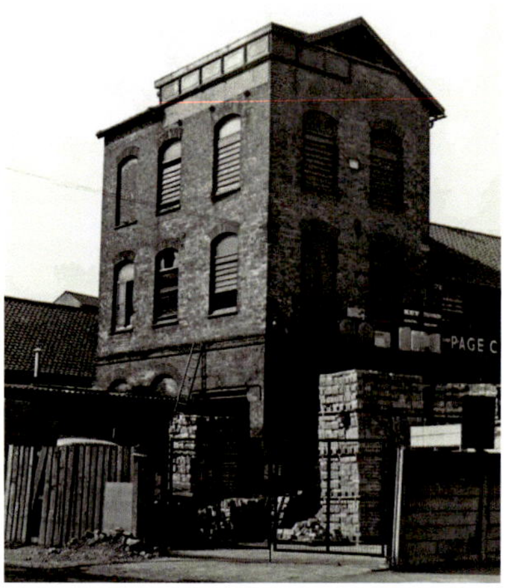

The Star Brewery, Ware, which went through four different owners.

Wickham owned or leased at least a dozen tied pubs and beerhouses in 1899. He was bottling his beer using equipment designed by his son, Dennis, who founded an engineering company that made everything from railcars to steam cleaning equipment.

William retired in 1899. The brewery and its tied houses were acquired by Holt's Marine Brewery of Ratcliffe, East London.

WATFORD

Benskin's

John Pope was born in Hemel Hempstead in around 1662. By 1693 he was a baker in Watford and had moved to premises in Shelton's Yard, off Back Lane (now New Street), by 1709. Some time between 1714 and the year he died, 1722, Pope started brewing at Shelton's Yard, which had become known as Pope's Yard.

After the death of Pope's son, Daniel, Pope's widow, Sarah, ran the brewery until she died in 1750, leaving it to her grandson, John, then twenty-one. His son, John II, expanded the business, and in 1812 he bought a large mansion in the Lower High Street, Watford, which had been built in 1775 by a successful brewer from Southwark called Edmund Dawson and had plenty of open space behind. Some time during the 1810s or 1820s John Dyson II completed building a new brewery in the Lower High Street, closing the original brewery in Pope's Yard in 1830.

John Dyson II died in December 1845, aged eighty-two, and his eldest son, John III, then thirty-six, inherited the brewery businesses, known by 1861 as the Cannon Brewery. In June 1867 John Dyson III died, aged fifty-eight. Five months later, the brewery, plus forty-two pubs and beerhouses, came up for sale at the London Auction Mart.

The successful purchaser, for a total of £34,000, was a fifty-three-year-old former hotel proprietor, Joseph Benskin, born in Barking, Essex, in July 1814. His great-grandfather, Richard Benskin, had been rector of Seagrave, Leicestershire, in the 1740s. His father, Thomas, was a schoolmaster, but became impoverished, and had moved to London by 1804.

Joseph Benskin married Sarah Parker in 1843, and they had two children. However, Sarah died, and in 1847 Joseph married Sarah's older sister, Maria. He ran pubs and taverns in London in the 1850s, but by 1859 he was in charge of the refreshment rooms at Swindon railway station. A year later he was landlord of the Queen's Royal Hotel by the station in Swindon.

Benskin's early partner in the Cannon Brewery was William George Bradley, a businessman from Swindon. (He is sometimes described as a Watford draper: this appears to be confusing him with Charles Percy Bradley, who ran a draper's business at 175 High Street.)

Just before Benskin died in 1877, aged sixty-three, he had transferred the business to his wife, Maria, now fifty-nine, who held a two-thirds share, and his eldest son by her, John Puzey Benskin, born in 1850. This may have been to avoid a tricky legal problem: according to Benskin family tradition, Joseph's marriage to Maria was illegal under Church of England

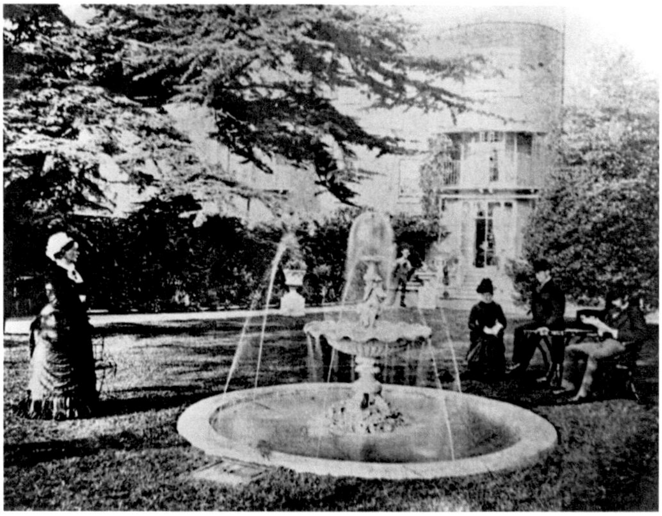

The Benskin family in the garden of their home alongside the brewery, *c.* 1885. That is, presumably, Maria Benskin on the left. (Courtesy of Favell Briggs)

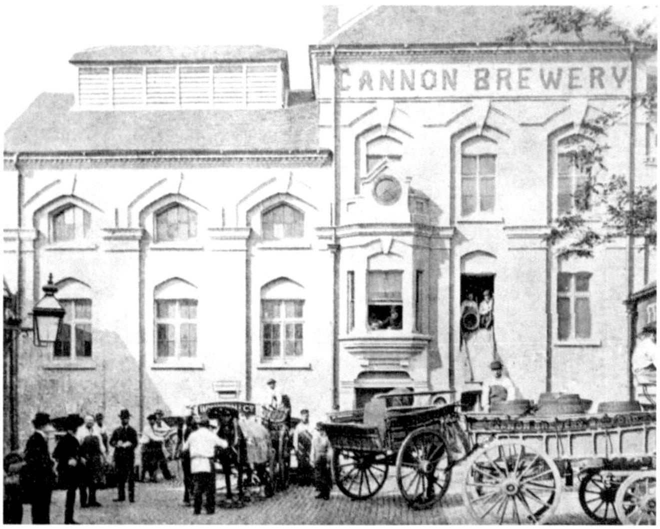

The Cannon Brewery, *c.* 1890. Several of the brewery workers are clearly wearing the traditional red stocking cap. (Courtesy of Watford Museum)

law because he had previously been married to her sister. This meant Maria's children were illegitimate and could not inherit Joseph's property. However, family tradition says that Joseph's son with Sarah, Joseph Clemson Benskin, was devoutly Christian and anti-alcohol, and renounced any interest in the brewery in favour of his half-brothers.

The brewery was enlarged again, under the brewery architect Arthur Kinder, in 1882. In 1891, Benskin's purchased the Bardfield Brewery in Great Bardfield, Essex. The company's cannon trademark was causing conflict with the Cannon Brewery in Clerkenwell, London. Thomas Benskin, Joseph's third son, had given his daughter, Doris, a brooch in the shape of a Thames lighterman's pennant, and she suggested a pennant would make a good company insignia.

A 150-quarter extension to the brewery was started in 1897. This inspired a wave of seven takeovers in 1897 and 1898, which added almost 300 pubs to the company's estate in less than a year: Groome's Brewery in Kings Langley; the Kingsbury Brewery, St Albans; Procter's Leighton Buzzard brewery; Healey's Brewery, Watford; Richard Ashdown's St George and Dragon Brewery in Leighton Buzzard; Down and Needham, a beer agency in Woburn Sands; and Hawkes and Co. of Bishop's Stortford.

Locke and Smith of Berkhampsted was acquired in 1911. The seventy or so pubs formerly belonging to Taylor's Brewery in Saffron Walden, Essex, taken over by Watney Combe Reid of London in 1899, were supplied from 1916, and then actually purchased in 1919.

In January 1920 Benskin's bought Pryor Reid's brewery in Hatfield and 105 pubs. Benskin's older neighbour in Watford High Street, M. A. Sedgwick & Co. of the Watford Brewery, was bought in 1923. The previous year Benskin's had snatched up Roberts and Wilson's Brewery in Ivinghoe, Bucks, with twenty-odd pubs. In 1929 it acquired another, Buckinghamshire brewer W. & G. Weller of Amersham, with 140 pubs.

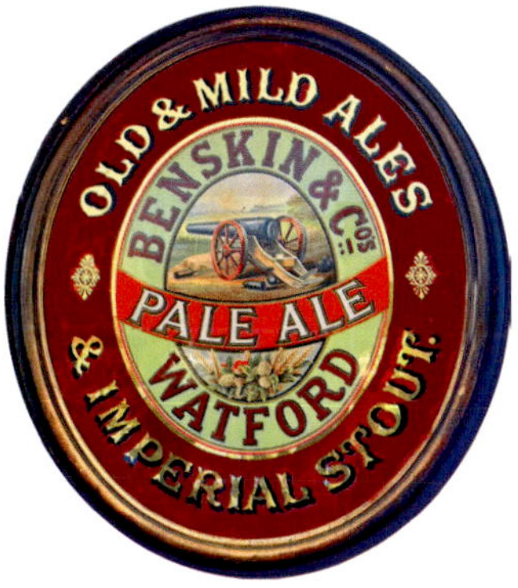

Right: A magnificent sign showing the cannon trademark, which Benskin's used before the now much more familiar pennant, on display in the Holly Bush pub, Hampstead.

Below: An illustration from the *Brewers' Journal* in May 1901 showing the new Benskin's Brewery. Topped by a sign displaying the brewery name and two cast-iron pennants, it was designed by George Adlam & Son of Bristol, assisted by the Watford architect Charles Ayres.

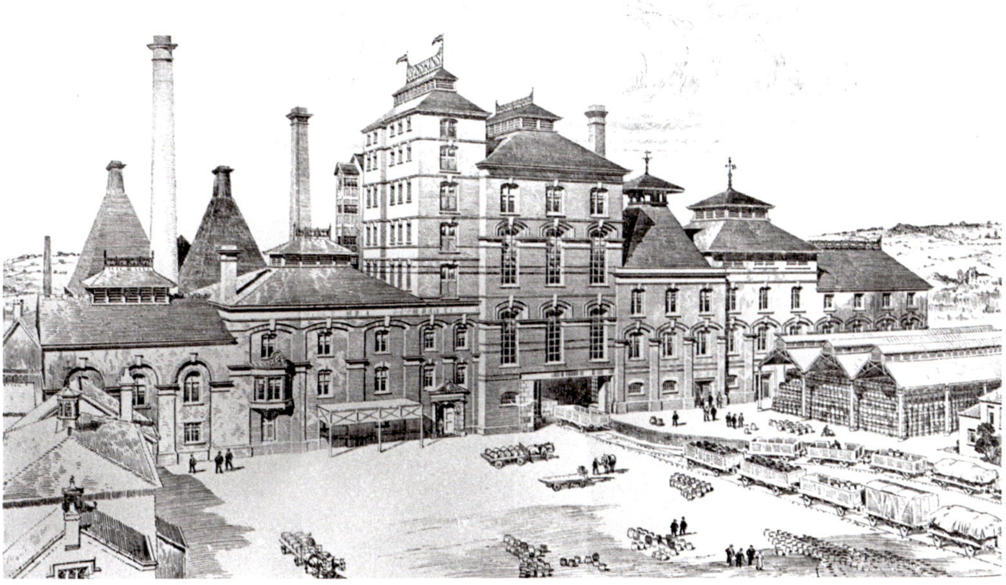

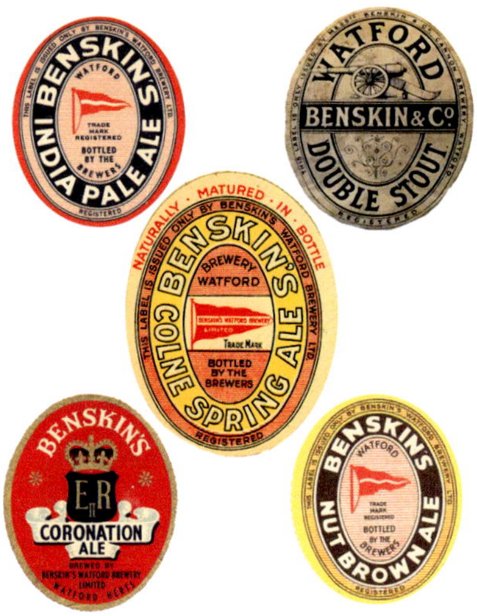

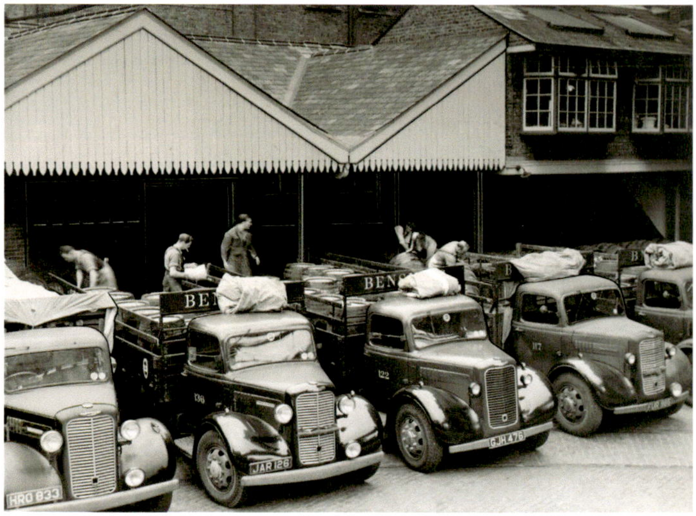

Above left: A selection of Benskin's labels, including one with the early cannon trademark and one for the famed Colne Spring Ale, a strong naturally conditioned beer that was aged for a year before bottling. The journalist and beer writer Alfred Barnard wrote in 1890: 'Never was there a more delicious beverage than this, full-flavoured, soft, creamy yet vigorous.' (Courtesy of Keith Osborne)

Above right: Commer drays, made in Luton, being loaded with wooden casks at the Benskin's Brewery, 1948.

Benskin's made its last takeover, of the tiny Wells Watford Brewery in St Albans Road, with a couple of dozen tied houses, in August 1951. By the mid-1950s Benskin's controlled 636 pubs and hotels and sixteen off-licences. Early in 1957 Ind Coope and Allsopp of Romford and Burton-upon-Trent swooped, its bid officially accepted on 13 March. Benskin's was now part of the biggest brewing concern in the country.

The Watford Brewery remained open. Then in 1961 Ind Coope merged with Tetley's of Yorkshire and Lancashire and Ansell's from Birmingham, and in 1963 this became Allied Breweries Ltd. Gradually Benskin's name disappeared, as did its beers. In 1972 the brewery was closed completely. The demolition men arrived in summer 1979, leaving only the old brewery offices – Edmund Dawson's mansion – still standing, now in use as Watford Museum.

The Benskin's name was revived by Allied Breweries in 1980, and a new Benskin's Bitter, brewed at Romford, appeared in several hundred pubs. That lasted until 2002.

Healey's Brewery, King Street

Charles Frederic Healey was one of four brothers, sons of George Healey, a Watford grocer, who all became owners of breweries in south Hertfordshire. Charles bought three plots in the sale of the Watford Place estate, King Street, in 1851. One held the estate stables, and these he converted and adapted into a brewhouse. Next door he built the New Inn.

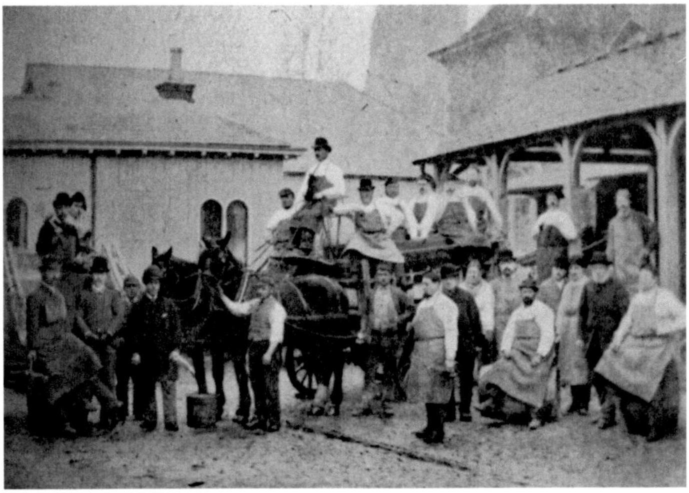

Healey's yard with the brewery workers in the standard stocking caps and aprons, 1883. (Courtesy of Watford Library)

He died in May 1863, aged forty-one. The business was carried on by his wife, Elizabeth Julie Healey, then thirty-three, and her son, Charles junior, born in 1856.

In around 1897 Healey's looks to have acquired the Victoria Brewery in St Albans Road. The following year, in January 1898, Healey's itself, and its fifteen tied houses, was taken over in the expansion of Benskin's Brewery. Charles Healey, generally known as Colonel Healey, joined the Benskin's board.

Victoria Brewery, St Albans Road

Samuel Roate, born in around 1822 in Hoddesdon, appears as a brewer in 'Newtown', Watford, in a directory of 1854. By 1866 his premises were named as the Victoria Brewery, Watford new town.

In February 1871 Roate was summonsed for being drunk and riotous in the public thoroughfare at 2 a.m. A month later the licence of the Victoria Brewery was transferred from Roate to Frederick Cocks.

Cocks was in charge of the brewery for twenty years. In July 1892 the Victoria Brewery was acquired by William Raymond Clarke, together with its off-licence.

Clarke remained at the brewery until at least 1896, but soon after it was acquired by Healey's Brewery. When Healey's was taken over by Benskin's, it continued to run the former Victoria Brewery off-licence as the Victoria Stores. It was later, by a strange coincidence, part of the Victoria Wine chain.

Watford Brewery

The Swan Inn, Watford, was mentioned as having a brewhouse attached in 1635. As the brewhouse at the Swan grew into a proper common brewery, it engulfed the original inn, which appears to have closed in around 1760, when it was owned by a miller, maltster and brewer called Samuel Smith. In 1859 the brewery was claiming a foundation date of 1750.

In 1790 the Smiths sold the brewery and its fifteen or so tied houses to George Whittingstall, a member of a rich family of millers, whose brothers owned the Hoddesdon Brewery and the St Paul's Brewery, Bedford.

Whittingstall died in 1822, aged around sixty-seven. Under his will, his fortune, and the property, were split between his sister, Elizabeth, and Edmund Fearnley, a thirty-five-year-old 'beast salesman' at Smithfield Market who was a great-grandson of one of Whittingstall's aunts. To inherit the whole property when Elizabeth died, George's will said that Edmund had to take the name Whittingstall.

Edmund Fearnley moved into the brewer's house in November 1822. Three years later Elizabeth Whittingstall died, and Edmund inherited the rest of the property, changing his surname by royal licence to Fearnley-Whittingstall. The brewery, however, continued to be known as 'Whittingstall's'.

Edmund died in March 1856. The reins were picked up by his only son, George Fearnley Whittingstall. He ran the brewery for six years, then leased it from 1862 to William Fellowes Sedgwick, a partner in a firm of Watford auctioneers. William died in February 1869, aged fifty-eight, and running of the business fell on his son, Frederick James, then twenty-two, and his wife, Mary Ann Sedgwick.

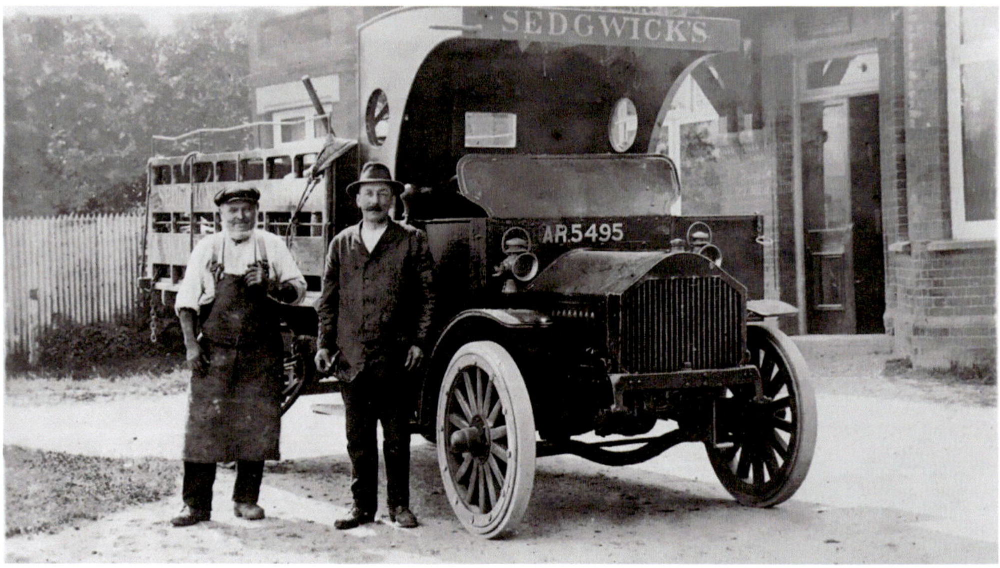

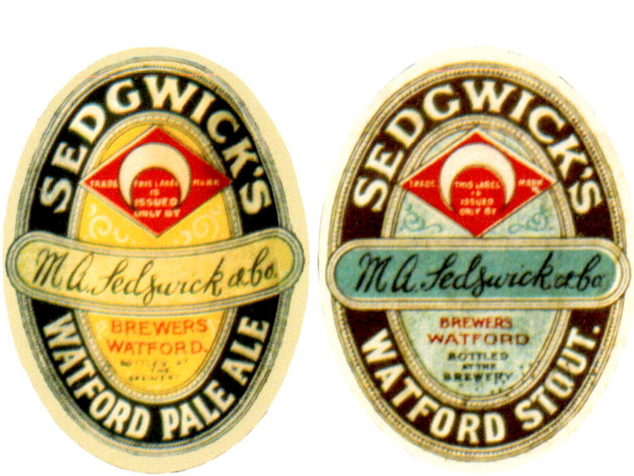

Above: A Sedgwick's motorised dray pictured outside an unknown pub between 1905 and 1920. Note the lack of windscreen and acetylene headlamps. Number plates bearing the AR prefix were issued in Hertfordshire from 1903 to 1921.

Left: A pair of labels from Sedgwick's Brewery showing the company's crescent trademark. (Courtesy of Keith Osborne)

In1878, Frederick's health broke. He eventually died twenty-one years later in an asylum in Northampton. His mother took over the running of the business, and in 1884 the name of the firm was changed to M. A. Sedgwick & Co. Around the time of Mary Ann's death in 1897, Sedgwick's bought Mercer and Sons' Colne Bridge Brewery in Uxbridge, Middlesex.

In October 1923, Sedgwick's and its ninety-seven tied houses were acquired by its rival, Benskin's, for £597,000, and the brewery was closed.

Wells & Co., Lion Brewery

Ralph Alfred Thorpe was born in 1855 in Wyberton, just south of Boston, Lincolnshire, the son of a farmer, Richard Wells Thorpe. His uncle, Thomas Wells Thorpe, ran the Eagle Brewery in Boston, one brother, John, became managing director of Daniel Thwaites & Co. of Blackburn in 1897, and another, also called Thomas Wells Thorpe, was a long-serving head brewer at Mann, Crossman and Paulin in Whitechapel and the inventor, in 1902, of Mann's Brown Ale.

In 1889 Ralph bought land in St Albans Road, Watford, which he then exchanged with the Paget Prize Plate Company for a more level plot. An artesian well was sunk, and the same year work started on building a six-quarter brewery on the site, to the design of the well-known brewery architect William Bradford.

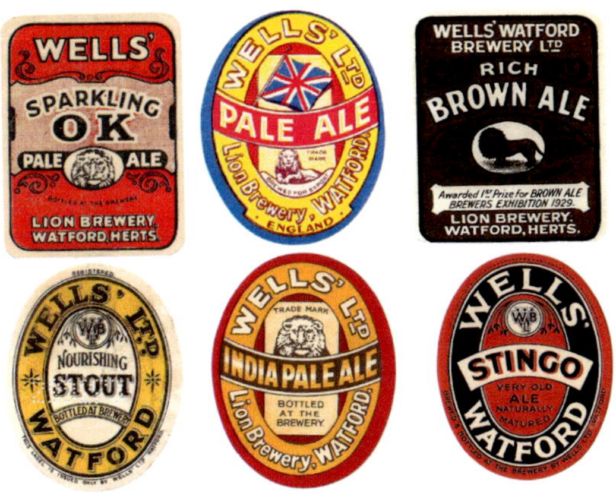

Above: A selection of Wells' Lion Brewery labels, including one for its answer to Benskin's Colne Spring Ale, Stingo. (Courtesy of Keith Osborne)

Right: An advertisement showing the Lion Brewery and its range of beers, *c*. 1906.

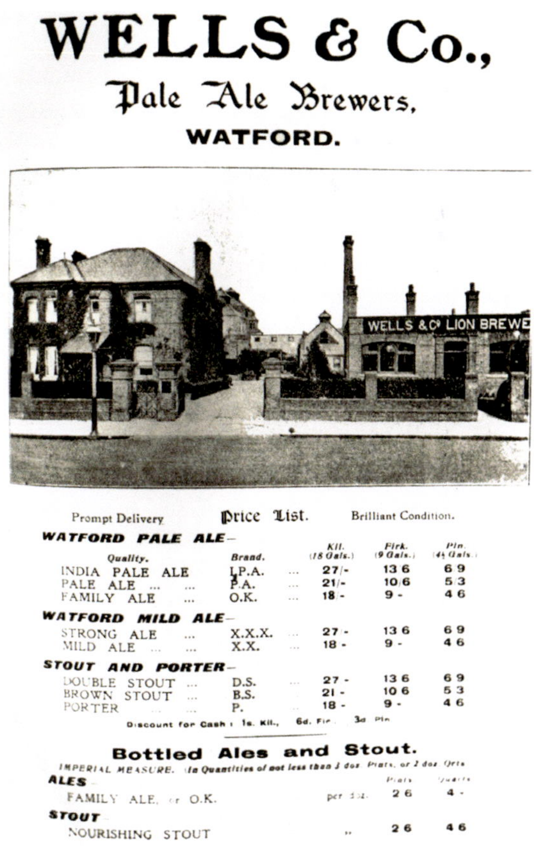

Ralph apparently picked one of his family's names, Wells, for the new concern, which was running by June 1890. In 1901 William Bradford was called in again to design a thirty-quarter brewery to run alongside the existing one.

The brewery amassed some sixteen tied houses and nine or so off-licences. It became a limited company, Wells Watford Brewery Ltd, in April 1925. By the start of 1951 it was clearly struggling, with no dividend paid on the ordinary shares for the year to March 1950. In June 1951 Benskin's stepped up and offered to buy the company for £1 a share, around a third more than their market value at the time.

WELWYN

The Welwyn Brewery

The Welwyn Brewery, behind Pound Cottages at the foot of School Lane, is first found in 1833, when it was being run by George Cass, who was also a corn and coal merchant. Cass is still listed as a brewer in Welwyn in 1839.

The census of 1851 shows James Deards, cooper, occupying the site. He had moved into brewing as well by February 1866. James died in August 1875, aged fifty-seven. The lease of the three-quarter brewery, together with a beerhouse at Woolmer Green, the Fox, passed to Samuel Gentle, for £30 a year.

Gentle was still at the brewery in March 1881. By 1882, it was in the occupation of Edward Kidd, born in Grantham, Lincolnshire, in around 1844. Kidd ran the brewery until September 1892, when the lease of the brewery went to Joseph Parker Mew, a partner in the Mew Langton Brewery in Newport, on the Isle of Wight.

Mew bought the Welwyn Brewery for his sons, Leonard Parker Mew and Joseph Harvis Parker Mew. They ran the Welwyn Brewery until November 1895, when the lease was acquired by Alfred Edward Macalister Hadwen, born in Lancashire in 1869. He stayed at the brewery for just fifteen months, before selling the lease of the brewery and the Fox to McMullen's of Hertford for £245.

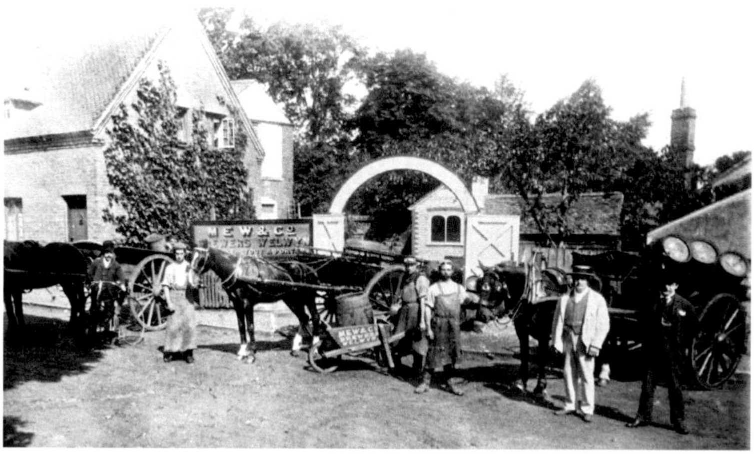

The Welwyn Brewery and its staff and workers pictured sometime between 1892 and 1895, when it was owned by the Mew family of the Isle of Wight.

WHEATHAMPSTEAD

Hope Brewery

James Wilkins, brewer of Wheathampstead, bought the Red Lion at Ayot Green in 1761. Twenty years later, in 1781, Wilkins was granted land by the Abbey of Westminster on Amwell Hill, Wheathampstead, to erect the brewery buildings that survive to this day and gave a name to the street on which they stand, Brewhouse Hill.

Wilkins died in April 1786. His son, James Wilkins junior, ran the business until 1803, when he looks to have sold it to Robert Parrott. Parrott was still supplying beer to Wheathampstead workhouse in 1825, but Robert Parrott, 'late of Wheathampstead', died in Harpenden in December 1830, aged seventy-nine.

What happened at the Brewhouse Hill Brewery in the 1830s and 1840s is unclear. In 1838 William Lattimore junior was paying rates on a brewery in Wheathampstead. This looks to be the brewery shown on the tithe-plot map of 1841 on the corner of the High Street and Marford Road. At some point William junior moved his brewery into the Brewhouse Hill premises, possibly in 1851.

William Lattimore died in 1864, aged fifty-one, and after his eldest son, William Henry Lattimore, died in 1875, aged thirty-one, the brewery was run by William Henry's mother, Ann.

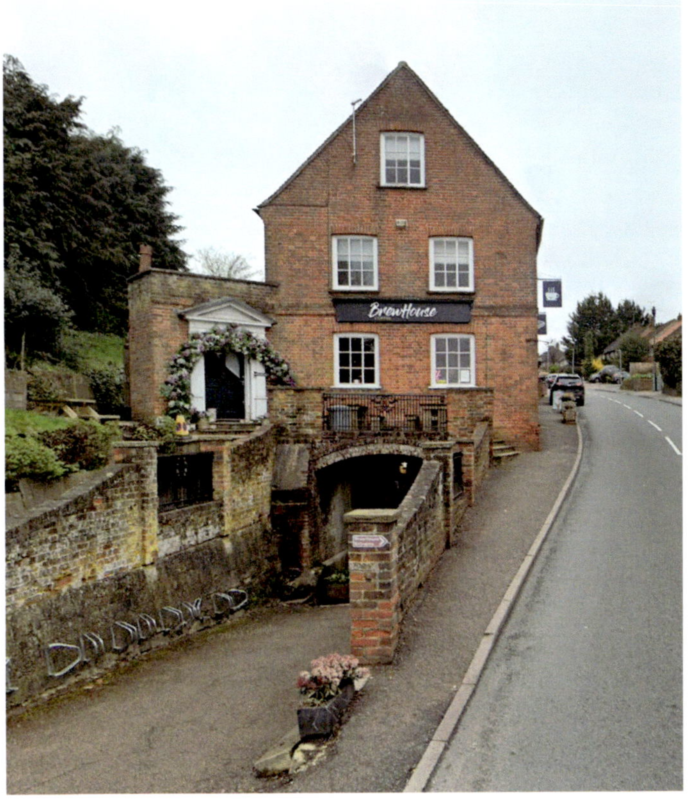

The former brewery on Brewhouse Hill, Wheathampstead, now brews a different drink – coffee, not beer. Under the building can be seen the cellars cut into the side of the hill that could be reached directly from the road.

Ann Lattimore died in July 1879, aged around sixty, and the business was continued by her younger son, Charles Frederick Lattimore, who was then around thirty-two. Charles died in May 1895, aged forty-nine. The following year, in April 1896, it was announced that 'the brewery of Messers Lattimore of Wheathampstead, with 26 public houses, has been sold to Messers Pryor Reid and Co.'. However, Pryor Reid only acquired around half the Lattimore tied estate, with other brewers also buying Lattimore pubs.

The brewery, meanwhile, was acquired by a firm called Wooldridge, Hill & Co., almost certainly the same company that ran the Bruce Grove Brewery in Tottenham.

Wooldridge Hill were there until 1902, when the five-quarter brewery, its twenty-quarter maltings further up Brewhouse Hill and eight tied houses were offered for auction at the London Auction Mart. The buyers were evidently a concern called Chambers & Co.

Not quite three years later, in November 1904, the Wheathampstead Brewery and its tied houses, including an off-licence, were again on the auction list at the Mart in Tokenhouse Yard. No buyer seems to have come forward, and the brewery shut.

Sutton

George Sutton and his son, George junior, who were coal and corn merchants and millers in Wheathampstead in the 1820s and 1830s, were listed as brewers as well in 1838, apparently operating out of the Mill House in the centre of the village.

In April 1841, however, the partnership between George Sutton senior and junior, brewers and coal merchants, was dissolved. It is very possible that the Suttons were running the Brewhouse Hill Brewery as the tithe-plot map of 1841 appears to show the Suttons owning the building on the brewery site.

WHITWELL

Archer

John Hill, a farmer at Bendish, is listed as a brewer in Whitwell in 1839. John died in April 1843, aged fifty-eight. He was followed at the farm and brewery by Robert Hill, born around 1791, who appears to be John's brother.

By 1850 Robert was in partnership as a brewer and maltster with his nephew, Frederick Hill Archer, born in Weston in around 1805. At some point a small brewery was opened on the north side of Whitwell High Street, and the 1861 census shows Edward Archer, evidently Frederick's half-brother, living there with his family and working as a brewer.

His place as brewer was taken by Frederick's son, Frederick William Archer, born in 1855. Soon after 1882 the brewery closed.

Appendix

Hertfordshire Breweries Founded Post-1980

Alford Arms

Frithsden

Whitbread's first attempt at a home-brew pub, opened in 1981. Closed in 1983, then reopened 1993–5.

Ash Valley Brewery

Prince of Wales, Green Tye, Much Hadham

A revival in 2014 by the pub's new owner, Rob Osborne, of the former Green Tye Brewery. Closed in 2018.

Banfield Ales

Unit 3a, Watermill Estate, Buntingford

Opened in May 2000. Subsequently, in September 2001, it shared its site with Woodlands Brewery, and the name was changed to Buntingford Brewery (q.v.).

Barley Brewery

Fox and Hounds, Barley

Opened by landlady Rita Nicholson in July 198. Closed in 1997 when the pub was bought by Inn Business.

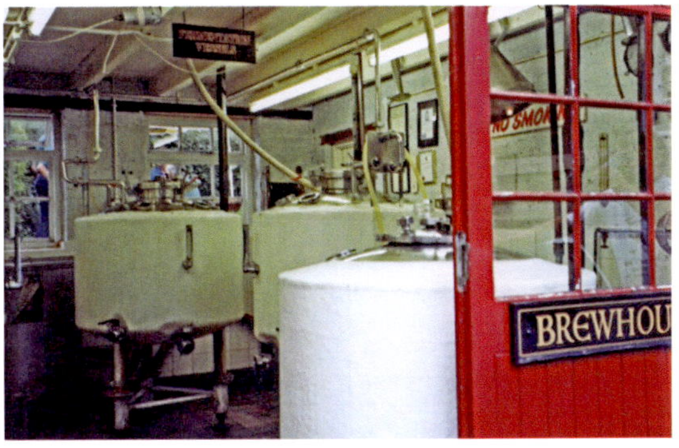

The brewhouse at the Alford Arms, 1994. The former cellar tanks have been repurposed as fermentation vessels. (Courtesy of Michael Croxford)

Baron Brewing

Brick House Farm, Great Hormead

Launched by Jack Baron in 2021. Still open in 2023.

The Belgian Brewer

Unit 11, The Links Business Centre, Raynham Road, Bishop's Stortford

Small brewery and taproom opened in 2018 by Nik Lemmens, producing beers brewed using traditional Belgian methods. Still open in 2023.

Berkhamsted Brewery

24a Wards Industrial Estate, Bourne End Lane, Bourne End

Demonstration brewery at premises of equipment manufacturer Inn Brewing, run by ex-Whitbread brewers Peter Shardlow and Robin Richards. Founded 1985, closed 1990.

Bowtie Brewers

78 Church Road, Watford

One-barrel brewery founded in 2018 in brewer Peter Hodge's home, with commercial brewing starting in 2019. Still brewing in 2023.

Bridgewater Arms

Bridgewater Arms, Little Gaddesden

Revival of brewing at a former nineteenth-century homebrew inn. Started by landlord Bill Woods in October 1981, closed in 1988.

Bubble Works Brew Co.

30–32 Hermitage Road, Hitchin

Brewing started in 2018 at the Beer Shop, part of the chain of off-licences founded in St Albans in 2012 by John Gudgin and Ben Hudson. Still open in 2023.

Bubble Works Brew Co.

26a Northgate End, Bishop's Stortford

A third Beer Shop opened in Bishop's Stortford in April 2023, and a second Bubble Works brewery was due to open in the summer of 2023.

Buntingford Brewery

Greys Brewhouse, Therfield Road, Royston

Began in 2001 as a joint venture between Banfield Ales and Woodlands Brewery in Buntingford. Brewing suspended in 2003 and restarted in Royston in September 2005. Still brewing in 2023.

Creative Juices Brewery

Woodoaks Farm, Denham Way, Maple Cross

Started in 2019 by Ben Janaway and two friends. Still open in 2023.

Crossover Blendery

Lannock Manor Farm, Hitchin Road, Weston

Began in 2020, and wild-ferments wort made by Elgood's Brewery in Wisbech.

Dark Horse

The White Horse, Castle Street, Hertford, 1994–6; Adam's Yard, Maidenhead Street, Hertford, 1996–2002

Founded by Ian Harvey, landlord of the White Horse.

Farr Brew

Founded by Nick Farr and Matt Elvidge in 2014 at Samuels Farm, Coleman Green Lane, Wheathampstead. Now runs six pubs. Due to move to new premises near Redbourn in 2024.

Fishery Inn

Fishery Inn, Watford Road, Elstree

Brewing started in October 1994 but stopped in 1995.

Flag & Firkin

Station Approach, Watford

Opened in 1997 as one of the Firkin chain of own-brew pubs, then owned by Allied Domecq, in the Pennant pub. Closed in 1999 when Allied sold the chain to Punch Taverns.

Fledgling & Firkin

Parliament Square, Hertford

Former Blackbirds pub, converted into one of the Firkin chain of own-brew pubs in 1996. Brewing stopped in 1999.

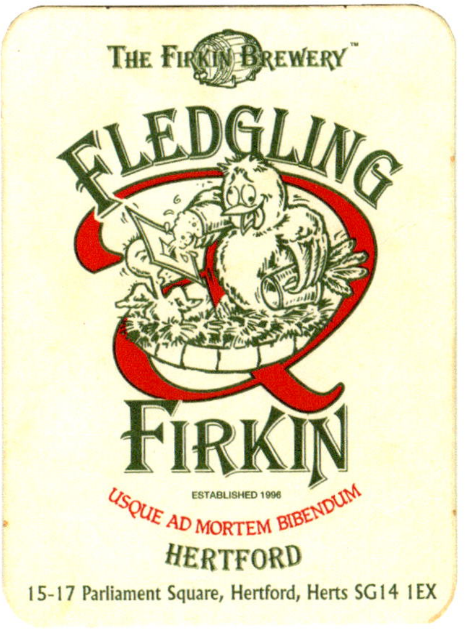

A beermat from the Fledgling & Firkin, one of three in the Firkin chain of home-brew pubs to open in Hertfordshire.

Foragers Brewery

The Verulam Arms, 41 Lower Dagnall Street, St Albans

Founded by Gerald Waldeck and George Fredenham in spring 2015. Brewing ceased in late 2019.

Garden City Brewery

22 The Wynd, Letchworth

Started in 2016 as a 'brewbar' in a former restaurant. Still open in 2023.

Garden Barber Brewery

41 Palmer Road, Bengeo

Run by landscape gardener Steve Johnson in the garden of his home. Opened in October 2001, closed early in 2004.

Green Tye

Prince of Wales, Green Tye, Much Hadham

Started by William Compton at the rear of the Prince of Wales pub in December 1999. Closed in 2013 but reopened the following year as the Ash Valley Brewery (q.v.).

Hadham Brewery

Hadham Industrial Estate, Church End, Little Hadham

Began brewing in 2015 with a ten-barrel plant using spring water sourced onsite. Closed in spring 2023.

Haresfoot

2 River Park, Billet Lane, Berkhamsted

Brewing started in March 2014, led by brewer Steve Carter. Moved to Preston Hill, Chesham, in 2017 but closed in around 2022.

Harpenden

Red Cow, Westfield Road, Harpenden

Started by Vivian Davies in August 1994. Closed in March 1997 when the two-barrel kit was moved to the Farmer's Boy in St Albans.

Lock 81 Brewery

Unit 21, The Wenta Business Centre, Colne Way, Watford

Opened by Henry Winter in 2019 with a one-barrel set-up. Brewing ceased in early 2023, though owners of the Wishful Drinking bar in Rickmansworth have beers brewed under the Lock 81 name by other breweries.

Mad Squirrel

Boxted Farm, Berkhamsted Road, Potten End

Started brewing as Red Squirrel Brewery in Mimram Road, Hertford, in March 2004. Moved to Potten End in 2011. Changed name to Mad Squirrel in February 2017. Still running in 2023.

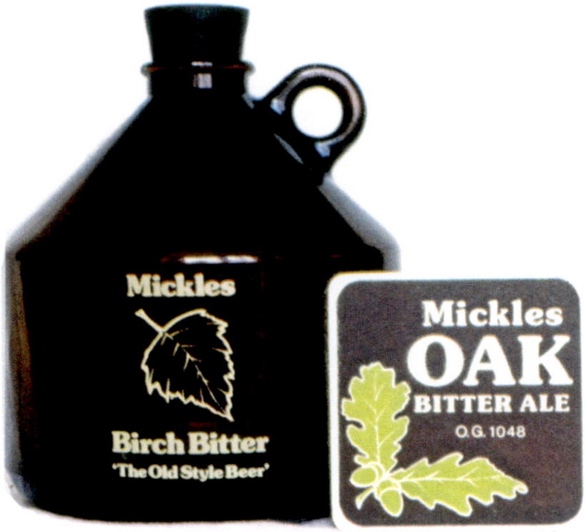

A 4-pint take-away plastic beer container, and a beermat, issued by the Mickles Brewery.

Mickles

At Dog Kennel Farm, Hitchin Road, Lilley, 1982–3, then the Victoria Brewery, Walkern (formerly Wright's), 1983–4

Founded by Mick Burch and Les Kent and called after their first names.

Mix

15–21 Market Street, Watford

A 'brew-it-yourself' brewery where brewer Mick Harrison helped customers brew their own commercial-quality beers. Ran from 1996 to 1998.

Mix

3 Cemmaes Court Road, Hemel Hempstead

Started in a domestic garage in September 2013 by Mick Harrison, fifteen years after his previous venture. Closed in 2019 and relocated to Blackfield, Hampshire.

New River Brewery

Hoddesdon Industrial Centre, Pindar Road, Hoddesdon

Started in July 2015, taking its name from the artificial river that brought water to North London, supplying several London breweries. Still open in 2023.

Old Cross Tavern Brewery

8 St Andrew Street, Hertford

The Old Cross Tavern was started in a former antiques shop in 1999. Brewing began in 2008 and ceased in 2018.

Original, Stevenage

Hollywood Bowl, Stevenage Leisure Park, Six Hills Way, Stevenage

First in a number of microbreweries opened by Bass in bowling alleys. Brewer was Steve Hibberd. Ran from 1996 to 2000.

Original, Watford

Hollywood Bowl, Woodside Leisure Park, Kingsway, Watford

Second brewery based in a Bass-owned bowling alley. Ran from 1996 to 2000.

Oxhey Village

Maxwell Rise, Watford

Started in 2016 by Chris Newstead, proper commercial brewing began in 2019. Still open in 2023.

Paradigm Brewery

Green End Business Centre, Church Lane, Sarratt

Started in January 2015 with a five-barrel plant. Closed on 31 August 2023.

Philanthropist and Firkin

11–13 Victoria Street, St Albans

Based in the former St Albans library, it was converted into a pub by the Firkin own-brew pub chain, and named after Andrew Carnegie, the American philanthropist who paid for thousands of libraries to be built, including in St Albans. Brewing started in 1994 and ended in 1999.

Pope's Yard Brewery

Paramount Industrial Estate, Sandown Road, Watford

Founded by Geoff Latham and his partner Barbara Leenen and named after the site of the brewery that eventually developed into Benskin's. Brewing began in autumn 2012, first at Geoff and Barbara's home, then at premises in Riverside Road, Watford, and next in the former Wemco building in Whippendell Road, Watford. In spring 2018 it moved to Frogmore Mill, Fourdrinier Way, Apsley. In summer 2023 the brewery moved back to Watford. Still open.

Sawbridgeworth Brewery

The Gate Inn, 81 London Road, Sawbridgeworth

Opened in July 2000 by Tom and Gary Barnett with the kit previously used at the Alford Arms. Closed early in 2019.

Six Hills Brewing

29B High Street, Stevenage

Started by Paul Clinton as Bog Brew Beers in 2017 to supply the restaurant he ran with his wife. Rebranded using the name of a local landmark in 2021. Still open in 2023.

Summershed Brewery

Brew Cottage, The Bit, Wiggington

Set up in the home of Simon Crichton in 2016 with the beers going on public sale in 2018. Rebuilt in 2019 but closed in 2020.

Swannells

Numbers Farm, Station Road, Kings Langley

Alan Swannell's second venture after the Victoria Brewery in Ware. Ran from November 1982 to late 1983.

3 Brewers of St Albans

Symonds Hyde Farm, Symonds Hyde Lane, Hatfield

Founded by Mark Fanner and Pete Zivkovic with head brewer Matt Sutton. First brewed in 2013. Uses water from own borehole. Still open in 2023.

Tring Brewery

Dunsley Farm, London Road, Tring

Founded in December 1992 at 81–82 Akeman Street, Tring, by Richard Shardlow, previously a brewer with Greene King and Devenish. Supplying 100 outlets by 1998. Moved to new site in 2010. Still going in 2023, it is Hertfordshire's second oldest brewery.

Two Bob Brewing

Danesbury Avenue, Welwyn

Founded by Rob Clarke and Bob Underwood in 2021. Still going in 2023.

Verulam

The Farmer's Boy, 134 London Road, St Albans

Started in March 1997 by Vivian Davies using plant previously at Harpenden Brewery. Closed in 2015.

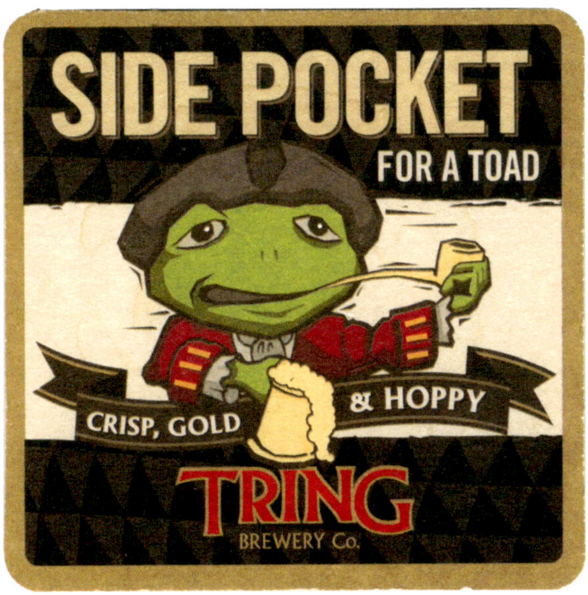

A beermat for Tring Brewery's Side Pocket for a Toad, a name derived from an old Hertfordshire expression meaning something not needed.

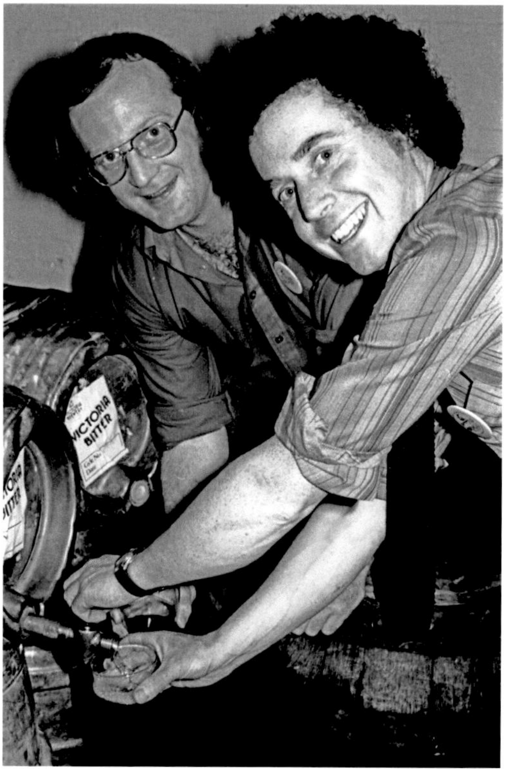

The late Tony Burns, left, and Alan Swannell, right, founders of the Victoria Brewery, enjoying a glass of their own beer. (Photograph by Nigel Willoughby)

Victoria Brewery

Victoria Maltings, Ware

Founded by Tony Burns and Alan Swannell in 1981, the first new brewery in Hertfordshire, in a disused maltings. Closed in 1985 after cash-flow problems.

Watling Street Beer/Radlett Brewery

Aldenham/Watford/Haringey

Started in January 2015 by Rudi Keyser in Aldenham. Relocated in 2016 to Hilfield Farm, Hilfield Lane. Later moved to an industrial estate in Greycaine Road, Watford, and then to the Triumph trading estate in Haringey, North London. Closed by June 2020.

White Hart Tap

4 Keyfield Terrace, St Albans

Originally the 'tap' of the White Hart inn on Holywell Hill, which had brewed its own ale in 1535. Brewing started at the Tap in late 2015 but ceased in autumn 2019.

Woodlands Brewery

27 Woodlands, Bishop's Stortford

Short-lived operation opened in 2001 and closed the same year after production was moved to the Buntingford Brewery.